Uses of Epoxy Resins

Uses of Epoxy Resins

W. G. POTTER

Chemical Publishing Company • New York

Uses of Epoxy Resins

ISBN: 978-0-8206-0202-8

Chemical Publishing Company:
www.chemical-publishing.com
www.chemicalpublishing.net

CHEMICAL PUBLISHING COMPANY, INC.

©1976, First American Edition

©2011, Second Impression

Printed in the United States of America

Preface

This book is, I believe, the first attempt to describe in a comprehensive way the techno-economic background to the use of epoxy resins. This task has not been easy, however, because of the vast number of different industries that use these resins, but to the author at least, this variety also makes epoxy resins a much more interesting group to study than many of the 'work-horse' thermoplastics.

This is not an exhaustive study of all uses for epoxy resins, and inevitably some uses have been omitted altogether. Again, because of limitations of space, it has not been possible to consider all important end-uses in the depth that perhaps they warrant. For example, automotive primers and can linings are given only brief consideration.

The book is addressed to the reader without an extensive chemical knowledge and will, I hope, be of interest to all who are concerned with materials or in cost savings. Throughout the book units of force are given in both lbf/in^2 and in SI units such as MN/m^2. The reader will recall that $1\ lbf/in^2 = 6.895\ kN/m^2$. In converting lbf/in^2 into SI units I have therefore adopted the approximation $1\ lbf/in^2 = 7\ kN/m^2$ and therefore $1000\ lbf/in^2 = 7\ MN/m^2$.

One difficulty in making a study of this nature is the lack of authoritative and detailed reports in the published literature on the economic considerations that underlie the use of epoxy resins in a particular situation. It has therefore been necessary to seek the views of resin manufacturers and end-users on the total set of considerations that lead to the choice of an epoxy composition for a particular job. I therefore wish to thank all of the many individuals and organisations that generously gave me their time. They are too numerous to mention individually, except that I particularly wish to record my gratitude to CIBA-Geigy for their help.

However, the book would not have been written at all without the generous help and cooperation of the Shell Group of Companies. I have been able freely to consult Shell International Research, and have had great assistance from both Shell International Chemical Company and Shell Chemical Company U.K. To all of these organisations and to the individuals in them I now express my thanks. In particular, I wish to thank Fred Parmenter for his continued encouragement and valuable suggestions.

Behind many authors there is often a long suffering family. To Brenda and to Sarah and Selina I wish to dedicate this book, for putting up with it for a further two years.

W.G.P.

Contents

Section I – GENERAL

1 Introduction 3
2 Why use Epoxy Resins? 15

Section II – SURFACE COATINGS

3 Anti-corrosive Paints 33
4 Marine Coatings 55
5 Coatings for Steel Pipes 64
6 Epoxy Powder Coatings 72

Section III – ELECTRICAL AND ELECTRONIC INSULATION

7 Rotating Machines 93
8 Distribution Switchgear and Transformers; Cable Jointing 107
9 Epoxy Resin Insulation for Outdoor Use 121
10 Electronic Applications; Printed Circuits 135

Section IV – CONSTRUCTION INDUSTRY

11 Floorings 157
12 Roads and Bridges 174
13 Adhesion and Grouting in Building 184
14 Grouts for Levelling; Miscellaneous Applications 199

Section V – OTHER PRINCIPAL APPLICATIONS

15 Composites 207
16 Adhesives 233
17 Arts and Crafts, Antiquities and the Home 249
18 Production Engineering 261

Index 277

SECTION I
General

1 Introduction

The amounts of epoxy resins manufactured and consumed are insignificant in comparison with polyethylene, polypropylene and polystyrene. Even when resins such as polyesters and polyurethanes are considered, the amounts of epoxy resins produced are smaller. But in terms of complexity of technology, variety and breadth of application, epoxy resins are surely superior to all other plastics and resins. There is hardly an industry in which these resins are not used. They find applications in electrical and electronic devices and oil wells, in space satellites and stained glass windows, on roads and bridges and in computers, in skis and in supersonic aricraft. They form the basis of high-performance paints which are used on ships, automobiles and as food can linings. They are used in the factory, the artist's studio, the laboratory and in the home.

Epoxy resins can be used as adhesives, sealing compounds, casting resins, dipping compounds, moulding powders, paints and varnishes, powder coatings, flooring and anti-skid surfacings and as the resin matrix in reinforced composites. When in these various forms they can be manipulated by hot or cold spraying, brushing, roller coating and all the other paint application methods, knifing, dipping, pouring, high- and low-pressure compression moulding and injection moulding. The reasons for this diversity of applications lie in the fundamental characteristics of the resins. Thus, during cure, no volatiles or other by-products are formed and volumetric shrinkage is very small – of the order of 1–2 per cent. The fully cured epoxy resin systems have the well-known properties of outstanding adhesion to many substrates, chemical resistance, toughness, mechanical strength and high electrical resistance. Many variations of these properties can be achieved by adjustments in the formulation used.

It is the aim of this book to give existing users of and newcomers to the resins, and all those with an interest in materials, a review of the wide range of uses of epoxy resins, and wherever possible to show why epoxy resins were chosen rather than another material. In some instances, new uses for the resins might be suggested to people who know them well, and others may be encouraged to experiment with epoxy resins where they have not used them before. On another level, the book might give an insight into the way in which an extremely versatile product can enter into and become accepted by a host of dissimilar industries.

BASIC EPOXY RESIN TECHNOLOGY

This section gives a brief, non-technical account of some of the basic concepts underlying epoxy resin technology and touches on the complexities of formulating epoxy systems. A more comprehensive and technical account of these matters has been given elsewhere by the author[1].

Epoxy resins, also known as epoxide or ethoxyline resins, contain the epoxy group which is the chief centre of their reactivity. When manufactured, the resins are either liquids or solids and contain, on average, two epoxy groups per molecule. In this physical state, the resins are thermoplastic, that is, they can be repeatedly softened by heat and hardened by cooling. In fact, some epoxy resins are almost pure organic chemicals.

However, the essential feature of epoxy resin technology is the conversion of the resin into a hard, infusible three-dimensional network in which the resin molecules are crosslinked together by means of strong covalent bonds. This conversion can be termed polymerisation, but is more commonly called curing or hardening of the resin. The reagents that bring about this change are known as curing agents or hardeners, these terms being completely interchangeable. It is in this cured form that the resins are almost always used; in the uncured, 'non-crosslinked' state they are of limited use.

Curing is an irreversible change and once the resin has been cured it cannot be recovered again in its original form. Cure can be slowed down, stopped or speeded up, but it cannot be reversed. Hence epoxy resins fall into the category of thermosetting polymers, which, once polymerised, cannot be re-used by melting and reprocessing. Continued heating of a thermoset merely leads to softening and eventually degradation and breakdown of the material.

In practice, cure is brought about simply by mixing the resin and curing agent, the two materials sometimes being dissolved in a suitable solvent. Once this mixing has occurred, a point of no return has been reached; cure begins and proceeds at a rate dependent upon factors such as temperature and the reactivities of the resin and curing agent towards each other. Some curing agents will react with the resins at room temperature or below, while others require heat to effect the polymerisation. The curing reaction is exothermic, that is, heat is evolved during the crosslinking process, which causes an increase in the temperature of the system. This is often called 'exotherm', and can reach high levels, perhaps 150–200 °C or above.

The temperature level reached in any particular example will depend not only upon the reactivity of the resin and curing agent, but also upon the temperature of the reactants and of their surroundings, that is, upon the rate at which polymerisation is occurring and the rate at which the heat evolved is being dissipated to the surroundings. Clearly, the ratio of the surface area to the mass of the reactants is important, as a large surface area would allow more heat to be dissipated compared with a smaller surface area for the same mass. This situation occurs if a mixture is being used as an adhesive rather than for casting purposes.

In practice, care is taken to ensure that the increase in temperature of a bulk mixture of resin and curing agent is not excessive. In the absence of such control, bubbling, cracking, charring and even complete degradation of the resin could occur in severe cases.

The time taken from the initial mixing of the resin and curing agent to the point when the viscosity of the mixture has become so high as to render the mix unusable is called the 'pot life' of the system. This time is therefore the practical working life of the mix, during which the material must be applied to the job concerned. To a large extent, the pot life is influenced by the same factors that affect the exotherm, and pot lives can vary from a few seconds to several weeks, such is the wide range of possibilities with epoxy resin formulations. The pot life is also dependent upon the intended application of the system; thus a viscous mixture may still be pourable into a mould but would not be suitable for glass-fibre impregnation in a laminating process.

The simple resin–curing agent combination alone seldom provides a material with all the properties required for use in a given application, and other materials must be added so as to modify the properties of the cured resin or to make it cheaper. The correct choice of the types and amounts of the different components of an epoxy formulation is a fairly precise, difficult and important task because the final properties and eventual performance of the system depend upon it. The various classes of materials that can be added to the resin and curing agent combination are:

(1) Diluents.
(2) Inert fillers and pigments.
(3) Flexibilisers.
(4) Fire retardants.
(5) Resinous modifiers.
(6) Cure accelerators.
(7) Reinforcements.

Some modifying materials frequently perform more than one function at the same time; thus a diluent might also be a flexibiliser or a cure accelerator. The classification does, however, provide a convenient basis for discussion.

Diluents

Diluents are liquids used to reduce the viscosity of the mixture. They can react chemically with the curing agent and resin, or be non-reactive, and are used in small amounts, generally of the order of 5–10 parts per 100 parts (w/w) of resin.

Inert fillers

Inert fillers have a variety of functions, but are chiefly used to make the system cheaper and to modify the physical and mechanical properties such as thermal conductivity and expansion, hardness and compressive strength. Other fillers are used to impart thixotropy to the mixture, that is, to thicken the mix so that it will not run off inclined surfaces but will flow when stirred.

Flexibilisers

Cured epoxy resins are usually hard but brittle solids. If a tougher, more flexible material is required, which is better able to withstand mechanical stressing or thermal shock, the epoxy system may be modified in one of two ways:

(a) Long, flexible molecular chains may be incorporated into the network during cure. This is achieved by using a flexible epoxy resin, curing agent or reactive additive.
(b) Long-chain molecules that remain unreacted after cure of the resin may be incorporated into the system. This effect, which is often regarded as plasticisation rather than flexibilisation, is achieved with non-reactive additives.

Fire retardants

Techniques employed to improve the fire-retardant properties of the cured resins mostly involve the incorporation of bromine or chlorine atoms into the system, usually by utilising halogenated epoxy resins or chlorinated curing agents. Certain fillers and diluents are also valuable in this respect.

Resinous modifiers

A wide range of other resins will react chemically with the epoxy resins to form a modified or 'alloyed' polymer system. Common examples are the combinations of epoxy resins with phenolic, urea melamine, polyester, polyurethane, silicone, vinyl and nylon resins. These systems have enhanced properties in certain respects over the properties of the individual, separate resins.

Other resins, although not reacting chemically with the epoxy resins, are also used to cheapen and alter the properties of the unmodified epoxy resins. Important examples are the addition of coal-tar pitch to yield epoxy paints with much improved water resistance, and petroleum-derived bitumens to cheapen anti-skid road surfacings.

Cure accelerators

Certain simple substances can increase the rate of reaction between the epoxy resins and some curing agents, and wide variations in the rate of cure can be achieved by the addition of small amounts of catalyst. For example, phenol causes the rate of reaction between an epoxy resin and a simple aliphatic amine to increase by several orders of magnitude.

Reinforcements

These are inert, fibrous materials whose prime purpose is to increase the strength of the system. Glass and metal fibres are widely used, the glass being in fibrillar form or woven into a cloth or mat and the metal fibres being used as discrete fibres or whiskers. Carbon fibres are also now becoming important. The whole subject of polymeric composite materials is based on the combination of a resin matrix with a fibrous reinforcement, and epoxy resins play an important part in this particular technology.

It can be seen, therefore, that what is generally referred to as an epoxy resin is more properly a carefully formulated mixture of ingredients. The correct formulation of these ingredients is a skilled task and much of epoxy resin technology consists in the successful development of formulations 'tailor made' for a particular application.

HISTORY OF THE DEVELOPMENT OF THE RESINS

Work on diepoxides was mentioned in patents in the late 1920s and early 1930s, but the beginning of today's epoxy resin technology is taken by many to be the patent of Schlack of IG Farben[2], whose application date was December, 1934. This patent was primarily concerned with the production of polyamines but mentions polyglycidyl ethers of diphenylolpropane (DPP). It does not, however, acknowledge these glycidyl ethers as resins. The first patent so to acknowledge the resinous characteristics of these substances appears to be that of Moss[3, 4], whose application date was December, 1937.

Work was also being carried out in Switzerland and in 1938 Castan[5], a Swiss chemist employed by the dental products manufacturers De Trey Frères of Zürich, filed a patent, to be followed by a second in 1943[6]. These two patents described the production of diglycidyl ethers and esters, including a resin based on DPP and ECH (epichlorohydrin), and polymerisation of these resins with acid anhydrides such as phthalic anhydride, and organic and inorganic bases including amines.

Attempts in the early 1940s to market Castan's products as casting resins for dental use failed, and the patents were subsequently licensed to CIBA A.G. of Basle (now CIBA-Geigy). At the Swiss Industries Fair in 1946, this company demonstrated the use of an epoxy resin adhesive, Araldite type I, to bond light alloys and at the same time offered samples of an epoxy casting resin to four Swiss electrical companies. This introduction of the resins to industry can be taken as the beginning of the commercial exploitation of these remarkable materials.

Parallel with this European activity, the paint company Devoe and Raynolds in the U.S.A. had been working with Shell Chemical Corporation to develop epoxy resins suitable for the surface coatings industry. This led to a long series of patents by Greenlee and co-workers of Devoe and Raynolds, the first being filed in September, 1943[7]. The work by Greenlee covered methods of preparing the higher molecular weight DPP–ECH resins[8–11] and important ways of modifying the resins or combining them with other materials to form surface coatings. Outstanding amongst these methods were the esterification of epoxy resins via both their hydroxyl and epoxide groups to form drying or non-drying resin esters[12] and the combination of epoxy resins with phenolic[13, 14] or amino resins[15, 16] to produce stoving finishes. While this basic work was being carried out, the Shell researchers were investigating a vast range of possible curing agents for the resins, monoepoxide compounds as reactive diluents and other possible modifications to the resin–curing agent combination [17–22]. Shell obtained licences to the Devoe and Raynolds patents and began to market a range of liquid and solid epoxy resins under the name Epon in the U.S.A. and Epikote in all other countries, principally to the surface coatings industry. These early resins were based on DPP and ECH and they continue to be the most important class of epoxy resins throughout the world.

In the late 1950s and early 1960s, other types of resin began to appear in the manufacturers' catalogues, including epoxidised novolacs and other polyfunctional epoxy resins, resins derived from halogenated DPP for flame retardancy and resins to impart flexibility to castings. Greenspan and co-workers of the Food Machinery

Corporation, U.S.A., reported on resins derived from the epoxidation of polyolefins and a range of these resins was brought to the market. The research teams of Union Carbide Corporation (Philips and co-workers) and CIBA, Basle (Batzer, Ernst, Fisch, Porrett *et al.*) investigated very many cycloaliphatic diepoxides, and market development was commenced with a small number of selected products.

The period from the mid-1960s to the early 1970s saw the continued development of new types of epoxy resins and curing agents. Increased interest was shown in the various types of cycloaliphatic resins, especially in view of their good anti-tracking properties and resistance to ultra-violet light. The glycidyl esters of the three phthalic acids, which also show these same attractive properties, were offered commercially, and development work continued on the high molecular weight linear phenoxy resins. The regular appearance in the literature of new curing agents continued unabated. One noteworthy event among the many was the marketing of a polymercaptan curing agent, which, when used with an accelerator, would cure resins at an acceptable rate at 0 °C.

Equally important, the exploitation of the outstanding properties of epoxy resin systems in an ever-growing number of uses has continued. It is difficult to select individual applications for particular mention, but the development of electrodeposited epoxy primers, epoxy powder coatings, filament-wound epoxy–glass circuit-breaker tubes, epoxy transfer moulding powders, epoxy flooring and adhesives all deserve special mention. Useful reviews, occasioned by the twentieth anniversary of the commercial introduction of epoxy resins, have also been published on the progress of epoxy technology, and mention a number of other applications which are dealt with in later chapters of this book.

Undoubtedly, the future remains bright for epoxy resins. Much has happened since the early days when CIBA and Shell, in the late 1940s and early 1950s, first began to manufacture and market these new and versatile materials. Many more manufacturers have since begun to produce the resins, and this trend will certainly continue. However, it is undoubtedly true that the continued development and exploitation of epoxy resins requires the resources of major world-wide companies, which will limit the number of manufacturers who are really able to play a significant part in this future activity.

COMMERCIAL BACKGROUND

Since the 1950s, epoxy resin production and sales have steadily increased, and in the last 10 years the world market has grown on average by at least 10–15 per cent per annum. At present, in the early 1970s, the total world-wide sales are about 150 000 tons, and probably one third of this total is accounted for by sales in the U.S.A., about a half in Europe, and much of the remaining amount in Japan. In the U.K., sales in 1972 were about 10 500 tons. At least 95 per cent of the epoxy resins sold is of the one basic type – the diglycidyl ethers of diphenylolpropane. These epoxy resins, manufactured from epichlorohydrin (ECH) and diphenylolpropane (DPP; also known as bisphenol A), have always been the 'work-horses' of the epoxy resin business, and no doubt will continue to be so for many years to come. One important reason for their initial popularity was that DPP was and probably still is the cheapest suitable aromatic dihydroxy compound available. The DPP–ECH resins also offer a very attractive range of properties which clearly have met customer needs over the last 25 years. The remaining 5 per cent of epoxy resins sold includes the cycloaliphatics, glycidyl esters and amines, epoxidised olefins and the specialised glycidyl ether resins such as the brominated types and the epoxy novolacs.

An analysis of sales, in 1965 and 1972, by end-use for the U.K. market is given in *Table 1.1* and shows that the surface coatings industry still dominates the market as the major consumer of the resins. Electrical and electronic uses have contracted although they are still major outlets. Other uses individually remain a relatively small proportion of the total tonnage, although they probably play a more important part in the overall earnings picture for the resins. This situation has not altered markedly over the last 15 years, and the continued flow of new developments, especially from Shell in the coatings field, will ensure that the pattern does not change rapidly in the near future.

Naturally, the end-use patterns of sales in other countries are not identical with that for the U.K., but in most major markets the surface coatings industry remains a major consumer of the resins. The resins are now manufactured in most industrialised countries of the world. In Western Europe alone there are at least fourteen different companies producing the resins, and in addition there are plants in most East European countries. The early pioneering companies of Shell, CIBA and Devoe and Raynolds have now been joined by Dow, Union Carbide, Bayer, BASF, Hoechst and many other large and small European, American and Japanese companies.

Table 1.1

EPOXY RESIN SALES IN THE U.K.; END-USE ANALYSIS
(AFTER *British Plastics*[23,24])

Application	1965		1972	
	Tons	*%*	*Tons*	*%*
Solid grades				
Surface Coatings	3000	50	5000	47
Electrical	900	15	1200	11
Liquid grades				
Surface Coatings	–	–	1200	11
Electrical	900	15	1000	10
Tooling	240	4	200	2
Adhesives	300	5	500	5
Laminates	240	4	300	3
Flooring	240	4	600	6
Miscellaneous	180	3	500	5
	6000	100	10 500	100

Most of the producers sell epoxy resins and a range of curing agents, giving the customer suggestions for starting-point formulations for their particular applications. However, many customers do not wish to compound the formulations themselves, and would prefer to purchase a ready formulated system that would meet their particular requirements. In order to meet this need, some manufacturers have offered ready compounded materials, but in addition formulator companies have been developed, especially in the U.S.A. and Japan, who can be likened to paint companies. Thus the epoxy resin can reach the ultimate end-user in three different ways, which can be shown schematically as:

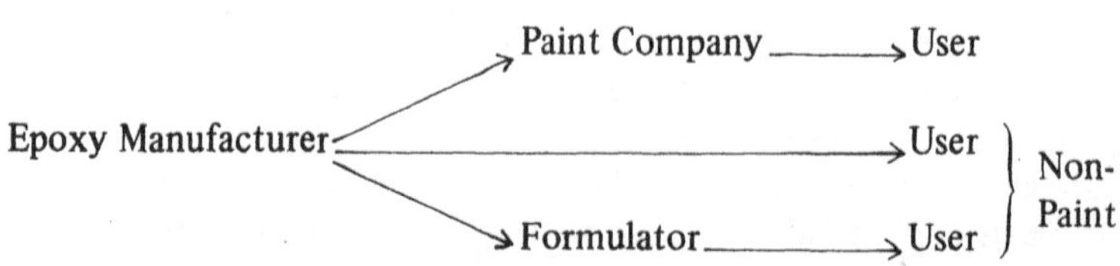

The formulators purchase the resins, curing agents and other raw materials and compound them into their own formulations, which are frequently packaged in convenient pre-weighed amounts in order to assist the customer. The composition of the formulations is usually kept secret. This, the formulator often feels, is his 'know-how' and security and he guards it well. The formulator not only provides a 'tailor-made' epoxy system but also supplies rapid technical assistance and advice to his customers. In fact, the real security of a formulator rests on the quality of the service he provides rather than on the secret constituents of his formulations. Large users of epoxy resins frequently consider whether it is cheaper to carry out their own formulating or to purchase from a formulator. There is no clear answer that meets all situations: favourable economics for one company may be unfavourable for another, and each case must be considered on its merits.

Epoxy resins are not cheap materials. Although prices vary between different countries, and are always subject to change at any time, the approximate level of prices in the U.K. in 1972 for liquid grades of the DPP–ECH resins was £600 per ton and for solid grades £550 per ton. Furthermore, epoxy resins will always be expensive, as they are made from expensive raw materials which require large capital-intensive plants for their economic production. The U.K. prices of curing agents in 1972 varied over a wide range, from about £500 per ton to as much as £2000 per ton for certain curing agents in the market development phase. Naturally, formulated epoxy systems can also show wide variations in price from below the price of resins to well above, and it would be meaningless to attempt to give any price indications for these materials. For comparison purposes, the 1972 U.K. prices for polyesters can be regarded as £200–250 per ton.

In the early days of epoxy resins, there were several key patents owned by the major resin-producing companies, which virtually dominated all of the applications for the resins. For example, curing with amines and acid anhydrides, two-part adhesives, esterification of the resins with fatty acids and the use of reactive diluents were all covered by patents. In addition, there were a number of basic manufacturing patents also possessed by a few companies, which helped to dissuade other manufacturers from entering the field. This situation has now changed. Many of the earlier patents have expired, and while new patents are still being added year by year, especially by the major manufacturers, no single end-use patent now enjoys the same domination as the esterification patents provided in the earlier days of epoxy resin development. Most of the new patents are concerned with novel curing agents or resins and rarely claim new principles or processes basic to the entire field.

REFERENCES

1. POTTER, W. G., *Epoxide Resins,* Butterworths, London (1970)
2. SCHLACK, P., *Ger. Pat.* 676 117
3. MOSS, W. H., *Brit. Pat.* 506 999
4. MOSS, W. H., *U.S. Pat.* 2 319 876
5. CASTAN, P., *Brit. Pat.* 518 057
6. CASTAN, P., *Brit. Pat.* 579 698
7. GREENLEE, S. O., *U.S. Pat.* 2 503 726
8. GREENLEE, S. O., *U.S. Pat.* 2 582 985
9. GREENLEE, S. O., *U.S. Pat.* 2 615 007
10. GREENLEE, S. O., *U.S. Pat.* 2 615 008
11. GREENLEE, S. O., *U.S. Pat.* 2 694 694
12. GREENLEE, S. O., *U.S. Pat.* 2 456 408
13. GREENLEE, S. O., *U.S. Pat.* 2 521 911
14. GREENLEE, S. O., *U.S. Pat.* 2 521 912
15. GREENLEE, S. O., *U.S. Pat.* 2 528 360
16. GREENLEE, S. O., *U.S. Pat.* 2 528 359
17. NEWEY, H. A., *U.S. Pat.* 2 553 718
18. NEWEY, H. A., *U.S. Pat.* 2 575 538
19. BRADLEY, T. F., *U.S. Pat.* 2 500 449
20. NEWEY, H. A. and WILES, Q. T., *U.S. Pat.* 2 528 932
21. WILES, Q. T., *U.S. Pat.* 2 528 933
22. WILES, Q. T., *U.S. Pat.* 2 528 934
23. *Br. Plast.,* 45, No. 1, 65 (1972)
24. *Br. Plast.,* 39, No. 1, 13 (1966)

2

Why use Epoxy Resins?

The title of this chapter asks a simple question to which there is no simple answer. For the designer and architect, the problem of material selection is often very complex and difficult. There is a bewildering variety of materials and processes available, and it is not easy to obtain a balanced and unbiased assessment of the merits and demerits of any individual material or group of materials from the information available. Load-bearing materials well illustrate the breadth of choice which confronts the design engineer. The range of suitable materials can extend from the traditional structural materials such as stone, wood, metals and concrete to the thermoplastic and thermosetting polymers, and to composite materials in which the stiffness and/or strength of a material is greatly enhanced by its combination with another material.

However, it is not the purpose of this book to consider the relative advantages of basic materials on such a broad scale. This has already been discussed by a number of authors elsewhere, for example by Alexander[1, 2] and Miller[3]. Rather, it is to consider why epoxy resins have been used for certain applications instead of another polymer or traditional material; what advantages epoxy resins have over other materials, and how these advantages can best be exploited. Plastics in general offer a combination of properties that is often superior to other materials in many respects. For example, plastics can have:

(1) The ability to be moulded directly into complex shapes.
(2) High electrical resistance.
(3) High corrosion and chemical resistance.
(4) Optical clarity.
(5) High strength to weight ratios.
(6) Attractive surface texture and colour.

In a similar way to metals, certain plastics can also be processed by mass production methods in which the product is made with consistent and reproducible accuracy. In contrast, large structures in glass-reinforced plastics are laid up by hand as single units.

Within these broad material characteristics, epoxy systems offer particular advantages as well as other additional attractions not yet considered.

COMPARATIVE ANALYSIS OF PROPERTIES

Physical properties

Plastics are comparatively light materials, their specific gravities being in the region of 1–2, which compares with values for metals of 2–12. Cast unfilled epoxy systems have specific gravities of 1.2–1.3, while some unidirectional glass fibre–epoxy laminates have values of up to 2. This light weight is a favourable factor when determining strength per unit mass of material.

The thermal properties of plastics are important from many points of view. In general, they have low thermal conductivities and can be considered as heat insulators. The thermal conductivities are in the range $2–10 \times 10^{-4}$ cal/cm/s/deg C and epoxy resins lie in the middle of this range, whereas metals are greater than 800×10^{-4}, wood 10×10^{-4} and cork 1×10^{-4} cal/cm/s/deg C. Such a low thermal conductivity can give rise to thermal gradients in operating equipment, and careful design is needed so as to avoid the formation of temperature high-spots which may lead to decomposition of the polymer and failure of the equipment. The thermal conductivity of epoxy resins can be increased by the incorporation of metal fillers and fibres into the system, but the enhanced values obtained still do not approach the values for metals.

Epoxy resins and organic polymers generally also have high coefficients of thermal expansion, some ten times greater than those of metals. This very large difference can present problems of differential thermal expansion and contraction when two dissimilar materials are used in contact with each other, for example when electrical and electronic epoxy castings and pottings contain metal embedments. Fillers can be used to reduce the coefficients, but again the values do not approach those of metals.

One major disadvantage of plastics is the limited temperature range in which they can be used, and especially the low upper temperature limit. This is a factor not normally encountered with metals or concrete, but it is a vital consideration when designing with plastics, for example when considering insulating materials for electrical machines. A knowledge of the highest and lowest temperatures to be experienced during storage, transport and use is absolutely essential when any article is to be fabricated from polymeric materials. *Table 2.1* gives an indication of the maximum long-term operating temperatures for some plastics materials. If very short exposures only are being considered, then a number of systems can withstand very much higher temperatures than the maxima indicated in *Table 2.1.* At present, certain epoxy systems are able to withstand long-term exposure to temperatures of the order of 210–260 °C. For example, one epoxy casting system showed a weight loss of 1.08 per cent after being kept at 210 °C for 500 h and a 9.20 per cent weight loss after 200 h at 260 °C[4]. Many other studies have been made of the high-temperature resistance of epoxy systems. In general, they have shown that in an oxidising atmosphere, certain amine-cured epoxy resins can start to degrade at 150 °C, whereas the anhydride-cured systems are stable at 200 °C[5]. Further, when an epoxy novolak resin cured with an anhydride was kept at 200 °C for 1000 h it lost only 2 per cent of its weight, whereas a regular DPP–ECH epoxy resin cured with the same anhydride lost 6.3 per cent of its weight under the same conditions[6]. Again, Lee[7] found that anhydride-cured DPP–ECH epoxy resins began to lose weight only at a temperature of 260–268 °C when heated at 9–10 deg C/min *in vacuo.*

Table 2.1

MAXIMUM OPERATING TEMPERATURES FOR SOME PLASTICS (AFTER MILLER[3])

Plastic	*Temperature, °C*
Phenolic moulding materials	150–250
Epoxy castings	50–260
PTFE	260
Polypropylene	120
Polycarbonate	120
Rigid PVC	90–100
Nylon 66	90
Polystyrene	65–85
Polyurethane (moulding grade)	88

Clearly, the performance of a polymeric system at any temperature depends critically upon the duration of its exposure at that temperature. An important advantage of epoxy systems is their ability to withstand high temperatures for longer periods than most other commercially available polymers, and this property is of great importance in certain applications, such as electrical machine insulation, printed circuit boards, and various aerospace uses.

Mention should be made at this point of the measurement of the heat deflection temperature (HDT). This is a widely used test that measures the temperature at which a standard bar of the cast resin system which is under a defined load undergoes an arbitrary deflection. The method of test is defined in ASTM D648-56, which requires a uniform stress of 264 p.s.i. to be exerted on a cast bar of dimensions 5 x ½ x ½ in, supported at each end and loaded in the middle. The bar is immersed in an oil-bath and the temperature of the bath is raised at the rate of 2 deg C/min. The HDT is taken as the bath temperature at the point when the deflection of the bar has reached 0.010 in.

HDT values for cured epoxy castings are in the range 50–200 °C although certain high functionality systems, such as those based on epoxy novolaks, can have HDTs up to about 300 °C if they are subjected to extremely long cures at high temperatures[4]. For comparison, HDTs of other polymers are given in *Table 2.2.*

Table 2.2

RANGE OF HEAT DEFLECTION TEMPERATURES AT 264 p.s.i. STRESS, FOR VARIOUS POLYMERS (SOURCE: *Modern Plastics Encyclopedia,* 1972)

Polymer	*HDT, °C*
Silicone moulding compound; glass fibre filled	480
Polyimide	360
Epoxy casting; aluminium filled	40–316
PF moulding powder; mica filled	150–260
6/6 Nylon casting	90–220
MF moulding powder	150–205
Polyester casting; rigid grade	160–205
Polycarbonate; filled	130–145
ABS	90–120
Polypropylene; filled	60–95

Care must be taken in using the HDT as a measure of the useful working temperature of a polymer system. The HDT is a single-point measurement and provides no information about the behaviour of a material at temperatures below the HDT when subject to different rates of heating. The HDT can be used only as a rough guide to the range of temperatures over which the various physical and electrical properties of the casting might be expected to remain constant or to vary very little. The preferred means of measuring the influence of temperature on the stability of an epoxy system is to measure a number of properties, such as flexural strength and dielectric constant, after exposing the casting to various temperatures for defined periods of time. The value of T_g (glass transition temperature) for the system is also a useful additional parameter.

The fire resistance of plastics is of importance to the designer. This property is frequently determined by the ASTM D635-56 test, which classifies materials as burning, non-burning or self-extinguishing. The test arbitrarily defines the burning rate in inches per minute, but does not specify a maximum elapsed time for a resin to extinguish itself. Current practice is to regard 15 s as the maximum time for satisfactory self-extinguishing properties. Normally, all cast epoxy systems will burn, but can be rendered self-extinguishing by the incorporation of a proportion of bromine-containing resins.

Mechanical properties

The very large number of different resins, curing agents, diluents, fillers and other modifiers available makes possible the preparation of cured epoxides with a wide range of properties. They can be made highly flexible and rubber-like, or hard and brittle, and they can have high or low heat deflection temperatures and be flame retardant. However, in general, an unfilled casting resin would have physical and mechanical properties that fall within the range given in *Table 2.3*.

A comparison of the mechanical properties of polymers and metals shows that, in general, polymers have a lower range of tensile strength and modulus of elasticity than metals and this lower rigidity compared with metals, concrete and stone is an important limitation to the wider use of polymers. However, if the specific stiffness (the stiffness to density ratio) is considered, the difference between plastics and other materials is greatly reduced and, in the case of the fibre-reinforced epoxy resins, the specific stiffness or rigidity exceeds that of metals, greatly so in the case of unidirectional, carbon fibre-reinforced epoxy resins. A similar situation is found when tensile strength is considered, both epoxy-glass and epoxy-carbon fibre composites having the highest strength to weight ratios of the structural materials[8].

Table 2.3

SOME GENERAL PHYSICAL AND MECHANICAL PROPERTIES OF AN UNFILLED CASTING EPOXY RESIN

Property	*Value*
Specific gravity	1.2–1.3
Thermal conductivity	4–5 x 10^{-4} cal/cm/s/deg C
Coefficient of thermal expansion	5–9 x 10^{-5} cm deg C/cm
Hardness	100–110 (Rockwell Mscale)
Ultimate tensile strength	4000–13 000 lbf/in^2
	28–91 MN/m^2
Elongation at break	1–3 per cent
Young's modulus	2–5 x 10^6 lbf/in^2
	1.4–3.5 GN/m^2
Ultimate compressive strength	15 000–30 000 lbf/in^2
	105–210 MN/m^2
Impact strength (notched Izod)	0.3–0.9 ft lb/in notch

A different approach to the comparison of structural materials was made by Alexander[1, 2], who compared the cost of '1 ton of strength' for various structural materials. Assuming an average manufactured cost of an epoxy-woven glass laminate of £4000/ton, the cost per ton per unit tensile strength was calculated to be 0.38 p/ton/in^2. On the same basis the cost of structural steel was 0.03, concrete 0.03–0.038, timber 0.05–0.07, copper castings 0.6–0.7 and brass strip 0.17–0.3 p/ton/in^2. Clearly, using these costs as a basis, there was no clear reason why an epoxy laminate, or any other polymeric system, should replace steel or reinforced concrete at that time (1967). The data used to arrive at these costs should naturally be brought up to date if this approach is to be used by a design engineer, but the results lead one to question why epoxy or polyester-glass laminated pipes, for example, have ever replaced steel pipes. Clearly, it is not sufficient for material selection to be based on one property alone. A cost-benefit analysis should be carried out on each property of the material and additionally on the other relevant aspects of the manufacture and use of the particular item under consideration.

Electrical properties

Nearly all plastics are good electrical insulators, but epoxy systems are especially valuable not only because of their very high resistivity, but also because of their very good physical and mechanical properties at normal and elevated temperatures, and their ease of fabrication into complex shapes. Electrical properties that are of importance in the choice of insulating materials and typical values for epoxy systems are:

(1) Volume resistivity .. .. $> 10^{16}$ ohm cm
(2) Surface resistivity .. .. $\sim 10^{13}$ ohm
(3) Dielectric strength .. .. 425 V/mil (0.125 in thick)
(4) Power factor .. .. 0.004–0.040 at 10^3 Hz and 23 °C
(5) Permittivity (dielectric constant) .. .. 3.0–4.5 at 10^3 Hz and 25 °C

These properties are not usually constant, but vary with time, frequency of applied voltage and temperature of the material. *Figures 2.1* and *2.2*[9] illustrate the variation in dielectric constant and loss factor

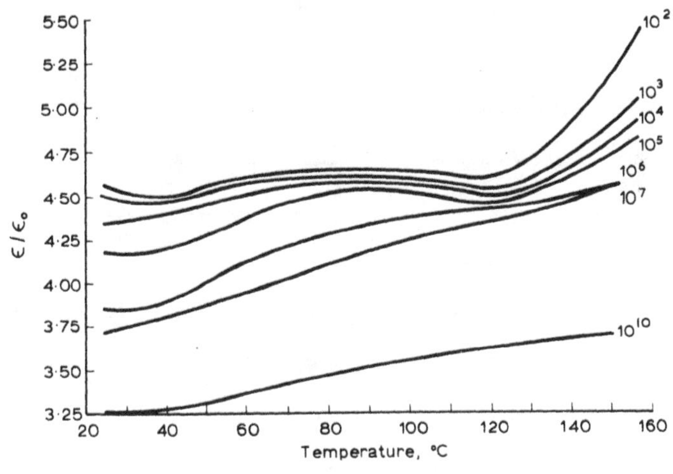

Figure 2.1. Variation of dielectric constant with temperature at frequencies from 10^2 to 10^{10} Hz for an epoxy resin (Courtesy Shell Chemical Company[9])

(power factor × dielectric constant) with frequency and temperature for one particular system, but this behaviour cannot be taken as an example of all epoxy systems and each one needs to be considered as a specific individual material in its own right.

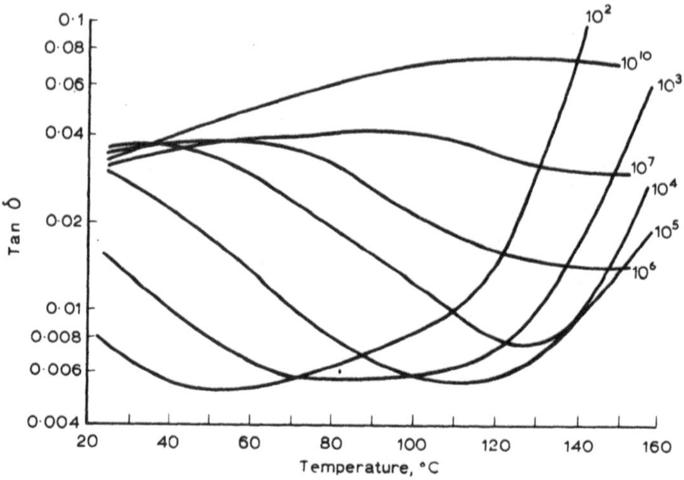

Figure 2.2. Variation of loss factor with temperature at frequencies from 10^2 to 10^{10} Hz for an epoxy resin (Courtesy Shell Chemical Company[9])

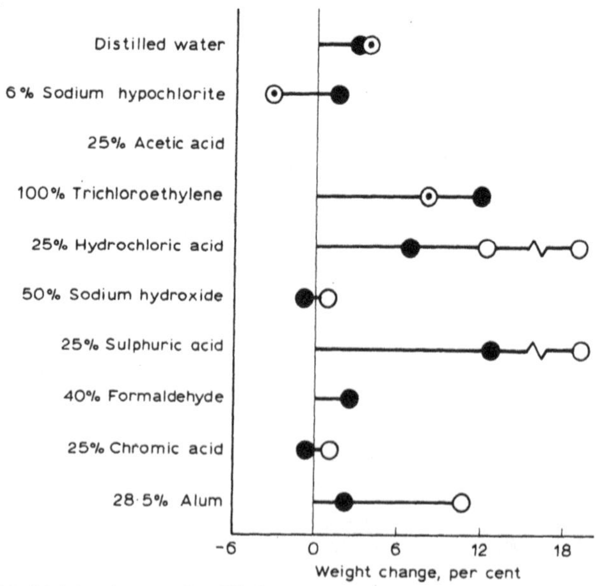

Figure 2.3. Weight change of unfilled epoxy resin–curing agent (TET) castings after 180 days' immersion: Curing agent TET (15 parts per 100 parts of resin); cured for 21 days at 25 °C. Immersion temperature: ● 24 °C; ⊙ 54 °C; ○ 82 °C. Total failure in acetic acid at 24–54 °C, 30–37 days; in formaldehyde at 82 °C, 90 days

Chemical properties

One of the outstanding characteristics of epoxy systems is their resistance to a wide range of chemicals, and many of the applications of epoxy resins utilise this property. Anti-corrosive coatings, monolithic industrial flooring and tank linings are all good examples of the use of the chemical resistance properties of epoxy resins.

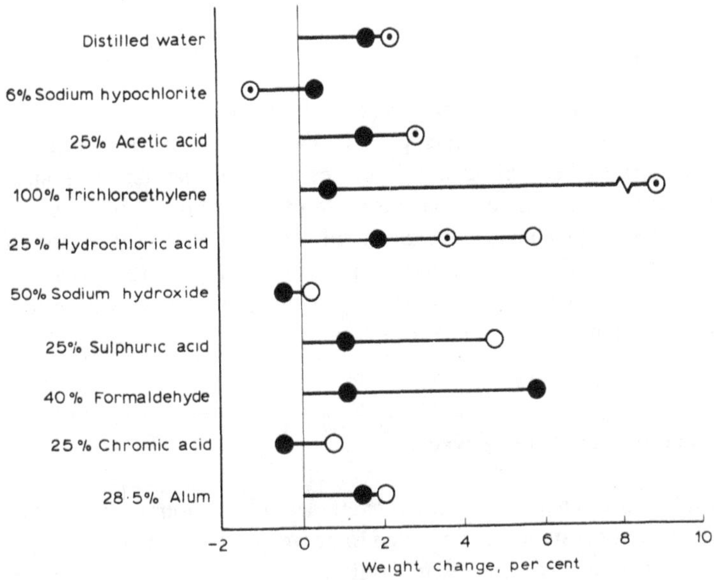

Figure 2.4. Weight change of unfilled epoxy resin–curing agent (MPD) castings after 180 days' immersion: Curing agent MPD (15 parts per 100 parts of resin); cured for 21 days at 85 °C plus 4 h at 150 °C. Immersion temperature: ● 24 °C; ⊙ 54 °C; ○ 82 °C

There are numerous examples and tables of data illustrating these characteristics. One method of measuring the chemical resistance of a casting is to determine the weight change or variation in mechanical properties after immersion in a reagent (ASTM test method D534-52T). *Figures 2.3* and *2.4* illustrate the chemical resistance, as measured by their percentage weight change, of unfilled castings prepared from a liquid epoxy resin cured with two different curing agents.

Processing properties

Epoxy resins are fabricated into their end-products by a large number of processes, which can, for convenience, be grouped into six different types:

(1) Coating.
(2) Bonding.
(3) Casting.
(4) Filament winding.
(5) Laminating.
(6) Moulding.

These methods are, of course, also used to manipulate other thermosets such as polyesters, phenolics and melamine resins, but epoxy resins offer significant advantages over the other resins because they react with their curing agents without the formation of water or volatile by-products, and exhibit only very low shrinkage during cure. Because of the wide range of resins and curing agents, it is also possible to arrange for epoxy systems to cure rapidly or slowly over the whole temperature range from 0 to 200 °C.

The properties of epoxy systems

In subsequent chapters, different examples will be given of the way in which epoxy systems have enabled improvements to be made in products or processes. These illustrations will, it is hoped, demonstrate the versatility of the resins and the advantages that can result from using them. The chief characteristics of epoxy resins are:

(1) The ability to be cured rapidly or slowly over a wide range of temperatures.
(2) The ability to be processed by a large number of different techniques.
(3) The absence of volatile by-products formed during the curing reaction.
(4) Low shrinkage during cure.
(5) Excellent adhesion to many different substrates.
(6) A high level of mechanical strength, which is retained at elevated temperatures.

(7) Outstanding toughness.
(8) Good electrical properties.
(9) Excellent chemical resistance.

REFERENCES

1. ALEXANDER, W. O., *Contemp. Phys.*, **8**, No. 1, 5 (1967)
2. ALEXANDER, W. O., *Futures*, **1**, No. 6, 500 (1969)
3. MILLER, W. G., *Br. Plast.*, **44**, No. 10, 93 (1971)
4. MEATH, A. R., in BRUIN, P. (Editor), *Epoxy Resin Technology*, Interscience, New York, Chapter 3 (1968)
5. POTTER, W. G., *Epoxide Resins*, Butterworths, London, 101 (1970)
6. LEMON, P. H. R. B., *Br. Plast.*, **36**, 336 (1963)
7. LEE, L. -H., *J. Polym. Sci., Part A*, **3**, 859 (1965)
8. HALL, M. M., *Br. Plast.*, **44**, No. 10, 85 (1971)
9. SHELL CHEMICAL CO., Technical Literature.

SECTION II
Surface Coatings

Introduction

For very many years, well over half of all the epoxy resins used world-wide has been employed as a film former in various paint systems. Even in 1973, the largest single application of epoxy resins remains in surface coatings, and epoxy resins have established themselves as an important group of film formers in their own right rather than secondary resins used to improve the properties of other materials. Some idea of the amounts of resin involved together with a comparative analysis of end-uses for epoxy resins has already been given in Chapter 1. The historical development of the use of these resins and the ways in which they are marketed, also described in Chapter 1, give some valuable insight into why the use of epoxy resin in paints has continued to dominate other applications. It is clear that a considerable investment in both manufacturing plant and laboratory facilities, together with marketing expertise, has ensured that continuing effort has been devoted to developing, manufacturing and marketing new epoxy resin systems and curing agents aimed at the paint industry and its customers, the end-users of the paints. The paint industry can now be regarded as one of the 'high technology' industries and is continually exploring new materials and application methods for surface coatings. In the industrial paints field, convincing evidence of the responsiveness and willingness to innovate by raw material manufacturers, paint manufacturers and paint users has been amply provided by the introduction and rapid adoption of, for example, the electrodeposition of paint, powder coatings, solventless coatings applied by hot airless spray techniques, and the wide use of epoxy-based coatings in the marine industry.

These lines of development have naturally been mostly dictated by economic factors, but health hazards and concern over pollution have become additional factors to be considered. Ever increasing costs in all countries have led to increasing pressure on paint manufacturers to produce the 'ideal' system which:

(a) can be applied at minimum cost in minimum time;
(b) has a low raw material cost;
(c) has a long life before breakdown under working conditions;
(d) provides the necessary protection to the substrate;
(e) is easily repainted; and
(f) does not contain toxic substances and does not contribute to environmental pollution.

The paint finally produced is inevitably a compromise between these desirable factors, but in particular (b) and (c) tend to be mutually contradictory. Another way of considering these factors is to look for 'minimum cost over the life of the object being coated'. Epoxy-based coatings have now been developed that meet many of these needs. High-solids and solventless coatings and powder coatings provide very thick films in one coat. Hot airless spraying also allows paint to cure at temperatures from 0 °C upwards, thus permitting painting to be carried out under a wide range of conditions. Wastage of paint is minimised by electrostatic spraying of powders or solvent-borne coatings. Epoxy–coal tar paints have been developed as a means of reducing raw material costs as well as improving certain performance properties. The use of organic solvents and their removal becomes unnecessary when powders, solventless coatings or paint applied by electro-painting are used.

While most surface coatings are used for both decoration and protection, it is possible to draw broad distinctions between coatings used primarily for decoration and those used primarily for protection. Epoxy resin-based paints are used overwhelmingly for protective purposes, although in protecting a surface its appearance is often also enhanced. The chief properties of epoxy resin coatings that have led to their widespread use are their outstanding chemical resistance, excellent adhesion to many different substrates, good physical properties such as flexibility and impact resistance and their ability to be formulated into paints that have different physical forms and require different techniques of application.

Paints must be brought to and spread on the surface to be painted and then the binder resin is polymerised so as to form a film. In order to be able to transport the paint in this way, it must be in a suitable physical form such as a free-flowing dispersion of solid pigment and filler in a solution of the resin, or a finely divided powder. In epoxy resin paints, the resin and curing agent or other copolymerising resin can be dissolved separately or together, usually in a concentrated solvent mixture. In solventless systems, the resin binder is usually a liquid that has a low viscosity when hot. Methods of applying epoxy paints can therefore be summarised as:

(1) Brushing.
(2) Spraying, including one- and two-component normal or heated components.
(3) Dipping, roller coating, flow coating.
(4) Electrophoretic deposition.
(5) Fluidised bed and electrostatic or flock-gun spraying for powders.

Once deposited on the surface the epoxy resin binder is polymerised to a tough film by the following chief methods[1]:

(1) The resin is cross-linked by polyamines or polyamides at ambient or slightly elevated temperatures.
(2) The resin is esterified by the resin or paint manufacturer through its epoxide and hydroxyl groups with suitable oil fatty acids to form an epoxy ester. Depending on the nature of the fatty acid, the resulting ester when applied to the surface being painted can either be polymerised by atmospheric oxygen at room temperature (an air-drying coating) or by heating it with or without a copolymerising resin (forced drying or stoving).
(3) The resin is blended with another resin, such as a phenol–formaldehyde resin, with which it copolymerises on heating (a stoving system).

These basic systems were developed historically for specific reasons, but all form coatings characterised by their excellent adhesion, chemical resistance, hardness, flexibility and toughness. The amine-cured system was the first to give stoved paint properties from a room-temperature curing paint. Combinations of an epoxy resin with other resins served to upgrade the performance of existing stoving systems, and esters provided a means of obtaining the performance of epoxy resin paints at a lower cost. Numerous reviews of the surface-coating applications of epoxy resins have been published, including those by Wheeler[2], Nickles[3], Maslow[4] and Gaynes[5].

REFERENCES

1. POTTER, W. G., *Epoxide Resins,* Butterworths, London, 145 (1970)
2. WHEELER, R. N., *Paint Technol.,* **18**, No. 207, 131 (1954)
3. NICKLES, O. L., *Off. Dig. Fed. Paint Varn. Prod. Clubs,* **27**, 185 (1955)
4. MASLOW, P., *Off. Dig. Fed. Paint Varn. Prod. Clubs,* **30**, 277 (1958)
5. GAYNES, N. I., *Ind. Finish. J.,* **35**, No. 8, 101 (1959)

3
Anti-corrosive Paints

In recent years, there have been several attempts to estimate the cost of metallic corrosion and protection in various countries. In the U.K. alone, it has been estimated[1] that this cost is about £1365 million per annum, or 3.5 per cent of the GNP. Other countries have also made rough estimates, for example, Sweden 1.25 per cent, The Netherlands 0.5 per cent and the U.S.A. 1.25 per cent of the GNP.

The estimated U.K. bill of £1365 million in 1971 is '. . . for preventing corrosion, compensating for it by overdesign, regularly inspecting its potential victims and rectifying some of its ravages'[2]. This sum, moreover, only deals with ten major industries and omits corrosion in agriculture and housing. It is also estimated that the annual U.K. cost of applying protective coatings is £772 million, of which £645 million is for organic coatings.

The growing awareness of the enormous sums of money that need to be spent in order to combat corrosion has also led to the use of sophisticated costing techniques to help determine which particular anti-corrosive system should be adopted in any particular situation. The results of these investigations have very often pointed to the use of epoxy resin-based paints, showing that the high initial cost of raw materials is more than offset by the greatly reduced frequency of maintenance required. In the painting operation, it has been estimated that the cost of surface preparation and paint application can be five times the cost of the paint for small articles and can rise to a factor of thirty or more for large structures such as bridges and ships[3].

Hence, rather than seeking economies in the cost of the paint, and possibly using a less effective coating, ways should be found of reducing the costs of preparation and application. The need for economies in this direction is also suggested by the diminishing availability of really skilled labour.

There is therefore a need for a paint system that can be easily applied at a minimum cost, and which will have a long maintenance-free life under operating conditions. There are many paint systems available that provide protection from the effects of corrosive environments, but most conventional paint films break down when severe conditions are encountered. Epoxy resin-based paints are able to withstand this attack because of their outstanding chemical resistance and physical properties and are therefore widely used in order to reduce maintenance and repainting costs to a minimum.

Three basic types of epoxy paints have been most used for anti-corrosive applications:

(1) Air-drying epoxy esters.
(2) Two-pack amine-cured systems.
(3) Epoxy–phenolic stoving systems.

The first two types have been in use for the last 20 years as general maintenance and protective coatings for all types of structures, whereas the stoved epoxy–phenolic systems have been used in a small number of selected applications where their extremely high chemical resistance was required and the articles to be coated could be heated to the necessary stoving temperature. This has led to the stoved epoxy–phenolic system being used chiefly in industrial and domestic appliance primers and finishes, storage drum linings and food can lacquers, collapsible tube internal lacquers and insulating varnishes.

Figure 3.1 shows diagrammatically the chemical resistance of three epoxy coatings: a two-pack amine-cured system, a solvent-free amine-cured system and a one-pack stoving system. The diagram suggests that with respect to many reagents, the chemical resistance of all three systems is similar at least up to 180 days' exposure.

Other commonly used maintenance coatings are the vinyl, chlorinated rubber and alkyd paints and a comparison of their chemical, physical and applicational properties as made by Scogin and Howard[5], is given in *Tables 3.1* and *3.2*.

The chemical resistance of the vinyls is about the same as that of the epoxy resins, but they show lower adhesion and heat resistance. The vinyls are also subject to attack by solvents, as they dry by solvent evaporation and can therefore be redissolved. Epoxy resins have applicational advantages over vinyls in that they can be applied as high-build coatings, which enables one-coat thick-film systems to be used.

Chlorinated rubber paints have outstanding resistance to alkalis and, to a certain extent, to mineral acids, but have lower solvent

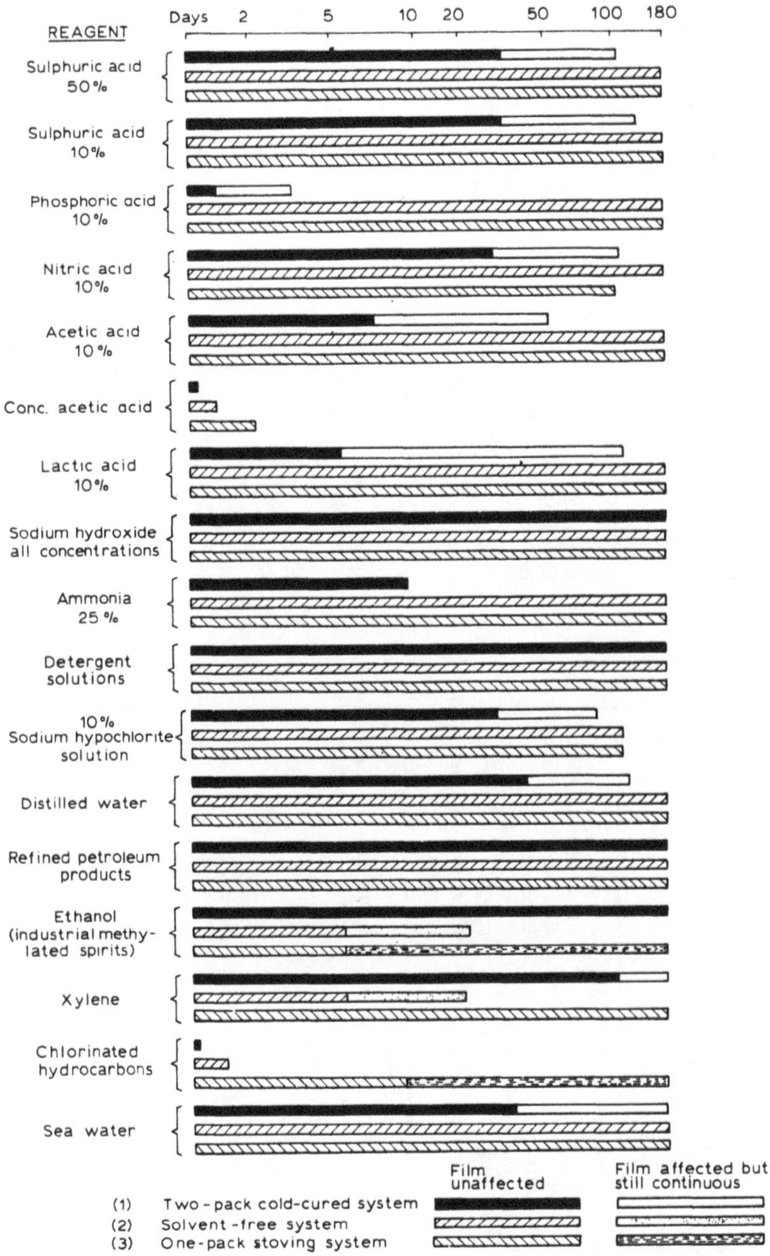

Figure 3.1. Chemical resistance of three typical finishes based on Araldite (CIBA-Geigy Ltd.) epoxy resins (Courtesy CIBA-Geigy Ltd.[4])

Table 3.1

COMPARATIVE CHEMICAL, PHYSICAL AND APPLICATION PROPERTIES OF EPOXY RESINS AND OTHER MAINTENANCE COATINGS (AFTER SCOGIN AND HOWARD[5])

Properties	*Epoxy amine*	*Epoxy ester*	*Vinyl*	*Chlorinated rubber*	*Alkyds*
Chemical resistance					
Acids	Excellent	Good	Excellent	Fair	Poor
Alkali	Excellent	Good	Excellent	Excellent	Poor
Solvents	Excellent	Good	Poor	Poor	Poor
Physical properties					
Flexibility	Good	Excellent	Excellent	Good	Good
Adhesion	Excellent	Excellent	Poor	Poor	Good
Hardness	Excellent	Excellent	Good	Good	Good
Abrasion resistance	Excellent	Excellent	Good	Good	Fair
Application properties					
Method	Spray, brush, roller	Spray, roller, flow, brush, dip	Spray, roller, brush, dip	Spray, roller, brush, dip	Spray, roller flow, brush, dip
Solids at spray viscosity					
Clear, per cent	40	35	20	25	40
Pigmented, per cent	55	50	25	30	50
Film thickness per coat, mil	1.0–2.0	1.0	0.4	0.5–0.6	0.75–1
Dry and/or cure	Chemical curing	Oxidation	Solvent evaporation	Solvent evaporation	Oxidation
Gloss retention	Poor	Fair	Excellent	Fair	Excellent
Colour retention	Fair	Good	Excellent	Poor	Excellent

Table 3.2

RESISTANCE PROPERTIES OF VARIOUS COATINGS* (AFTER SCOGIN AND HOWARD[5])

Resistance to	*Chemical-resistant vinyl*		*Alkali-resistant chlorinated rubber*		*Epoxy polyamide*		*Acid and fume phenolic*	
	Appearance	*Adhesion*	*Appearance*	*Adhesion*	*Appearance*	*Adhesion*	*Appearance*	*Adhesion*
Water soak corrosion	Excellent	Fair	Good	Fair	Excellent	Excellent	Excellent	Excellent
10% Sodium hydroxide solution	Excellent	Good	Heavy blisters	Poor	Slight discolouring	Excellent	Poor	Poor
10% Hydrochloric acid	Excellent	Poor	Excellent	Poor	Slight discolouring	Excellent	Slight blisters	Poor
10% Sulphuric acid	Excellent	Good	Medium blisters	Poor	Excellent	Excellent	Slight discolouring	Good
10% Chromic acid	Excellent	Poor	Very poor	Poor	Bad discolouring	Good	Poor	Poor
10% Acetic acid	Excellent	Poor	Excellent	Poor	Heavy blisters	Poor	Slight blisters	Poor
Glacial acetic acid	Slight blisters	Poor	Medium blisters	Poor	Heavy blisters	Poor	Medium blisters	Poor
Heat resistance	Complete failure		Complete failure		Discolours, very hard, slightly brittle		Discolours, brittle	
Abrasion resistance †	7.5		3		8		5	

* Test experience of a large chemical company.
† On a scale of 10.

resistance and adhesion compared with epoxy resins. The alkyd coatings possess the poorest chemical resistance of the three types.

Considering the two basic types of epoxy paint used for general maintenance and corrosion protection, in general, it can be said that:

(1) air-drying epoxy esters are used for mildly corrosive conditions;
(2) two-component amine-cured systems are used for highly corrosive conditions or where spillage of chemicals on to the paint will occur. In addition, epoxy powders coatings are being increasingly used to protect metallic surfaces against corrosion.

EPOXY ESTERS

The use of air-drying epoxy ester paints as primers and top coats for industrial plant maintenance under mildly corrosive conditions is now well established. Long-oil linseed esters are the most commonly used, frequently pigmented with red iron oxide for primer applications. This can be followed after 1–2 days by the same ester as a pigmented top coat or by an amine-cured paint if conditions are more demanding.

A characteristic of epoxy ester coatings is the fairly rapid loss of gloss when they are subjected to exterior exposure, owing to chalking of the film, and is particularly marked with short-oil esters. This phenomenon, which occurs to some extent when all DPP–ECH epoxy paints are exposed outdoors for long periods, is not due to the film breaking down and, moreover, epoxy ester films do not exhibit micro-cracking as do most alkyds.

Chalking can be claimed to be advantageous in that dirt cannot build up on the paint, the chalked film being self-cleaning. This can be regarded as a point in favour of epoxy esters and enhances their attractions for use as maintenance paints in refineries and petrochemical plants, except when the greater protection of amine-cured systems is required. Frequently, the ester is used pigmented with aluminium, which gives better gloss retention and good self-cleaning characteristics. Chalking is so slow in epoxy ester films that even after 5 years the decrease in film thickness can hardly be measured, and there is no loss of protection of the substrate over that period. In fact, experience indicates that the self-cleaning nature of the esters can extend the maintenance-free period of, for example, an oil storage tank by up to 14 years.

AMINE-CURED SYSTEMS

The wide use of amine-cured epoxy paints is mostly due to their providing many of the properties of stoved epoxy films from an ambient temperature-cured system. Their hardness, adhesion and flexibility on such diverse substrates as wood, ferrous and non-ferrous metals and 'green' and weathered concrete is outstanding and, while not offering such a high level of chemical resistance as the stoved phenolic and urea resin-modified epoxy paints, they exhibit better chemical resistance than most other air-drying paints.

All of the amine-cured paints have been extensively tested on exposure racks and in actual service. *Table 3.3* gives test rack-exposure results for a number of these paints that have been outdoors for 6–8½ years. Also included are results for a stoved epoxy paint and an epoxy-modified alkyd. Apart from the latter system, the film integrity is of a very high order in all instances. The general appearance of the film and, in particular, gloss, on the other hand, deteriorated owing to chalking of the paint.

The application of conventional solvent-borne epoxy coatings is labour intensive, as with other conventional types of protective coatings such as the vinyls and chlorinated rubber paints. Considerable efforts have therefore been devoted by resin manufacturers to develop epoxy coatings which allow thicker coats to be applied in one application. In this way, it is hoped to reduce the applicational costs of epoxy systems and improve still further their competitive position with respect to other paints.

HIGH SOLIDS AND SOLVENTLESS COATINGS

As labour costs rise steadily, there is a growing incentive to reduce the time and labour involved in applying a paint system without reducing its capacity to protect and decorate the substrate. In the shipbuilding industry, for example, the increasing size of the areas to be coated and newer methods of building also require faster painting. In Japan, the block construction method now represents 70–80 per cent of all major cargo vessel building and the painting of the sections before they are moved on to the shipway is a bottle-neck.

The need is therefore to increase the dry film build per coat, which can be achieved by using paints that contain only a small

Table 3.3

OUTDOOR EXPOSURE RESULTS FOR DIFFERENT EPOXY PAINTS (AFTER SHELL INTERNATIONAL CHEMICAL COMPANY[6])

Paint based on	*Exterior exposure at Egham, England,* years	*No. of coats*	*Film integrity**	*General film appearance**
Epikote † 1001–amine adduct	8½	One of primer plus two of finish	9	4
Epikote 1001–amine adduct	8½	One of primer plus three of finish	9	4
Epikote 1001–Versamid ‡ 115 (1 : 1)	8½	One of primer plus two of finish	9	4
Epikote 1001–Versamid 115 (3 : 1)	8½	One of primer plus two of finish	10	4
Epikote 815–curing agent H-2	6	One	9	5
Epikote 815–Synolide§ 960	6	One	9	4
Epikote 1001–amine adduct–coal tar–aluminium	6	One	10	2
Epikote 1001–MF (stoved coating)	7½	Three	9	4
Epikote-modified alkyd	8½	One of primer plus two of finish	6	4

*Key: 10 = excellent; 0 = very poor.
†Epikote is a Shell registered trade name for a series of epoxy resins.
‡General Mills Inc., U.S.A.
§Cray Valley Products Ltd., England.

amount of solvent (often called high-solids coatings) or contain no solvent (solventless coatings). Other advantages are also gained from using these paints: fire risks are reduced as a result of the complete or near absence of solvents, and the need for extensive ventilation equipment to remove solvents is eliminated, as is the pollution of the atmosphere by the solvents. Solventless systems can also be made that cure at much lower temperatures than conventional solvent-containing systems (which are usually based on solid epoxy resins)[7].

These high-build systems can be applied at thicknesses of 150–200 μm per coat by conventional spraying or brushing methods, the film, and some ventilation is therefore still required. The presence if it is to be sprayed through normal one-component equipment. The short pot lives of these paints also limit the amount that can be mixed at any one time. Large amounts would undergo rapid viscosity build-up, which would occur at an increasing rate as the temperature of the mix increased owing to the curing reaction. The film thickness is also limited by the need for the small amount of solvent to escape from the film, and some ventilation is therefore still required. The presence of solvents may also reduce film performance owing to small amounts inevitably being trapped in the film. Thus, although good protection against corrosion can be achieved on well prepared steel by a single 200-μm thick film of a solvent-borne coating, there is a risk that low-temperature conditions during application and cure, or lack of adequate ventilation, can lead to blistering on immersion.

Despite these possible drawbacks of solvent-containing coatings, very good service can and has been obtained from these systems under many exacting conditions. Nevertheless, in order to overcome these difficulties, epoxy paint systems have now been developed that contain no solvents. In formulating such a solventless system, a number of problems must be overcome, particularly when good curing characteristics at low temperatures are required. These problems are concerned with the different reactivities required in the paint when mixed ready for use, and when it is in the film applied to the substrate. Ideally, low reactivity is preferred for the bulk system (i.e. long pot life) and high reactivity when the paint is curing in place on the substrate. These two requirements are virtually impossible to attain in one paint, and a compromise is therefore reached. In practice, short pot lives are accepted in order to obtain a system that will cure at temperatures in the 0–10 °C

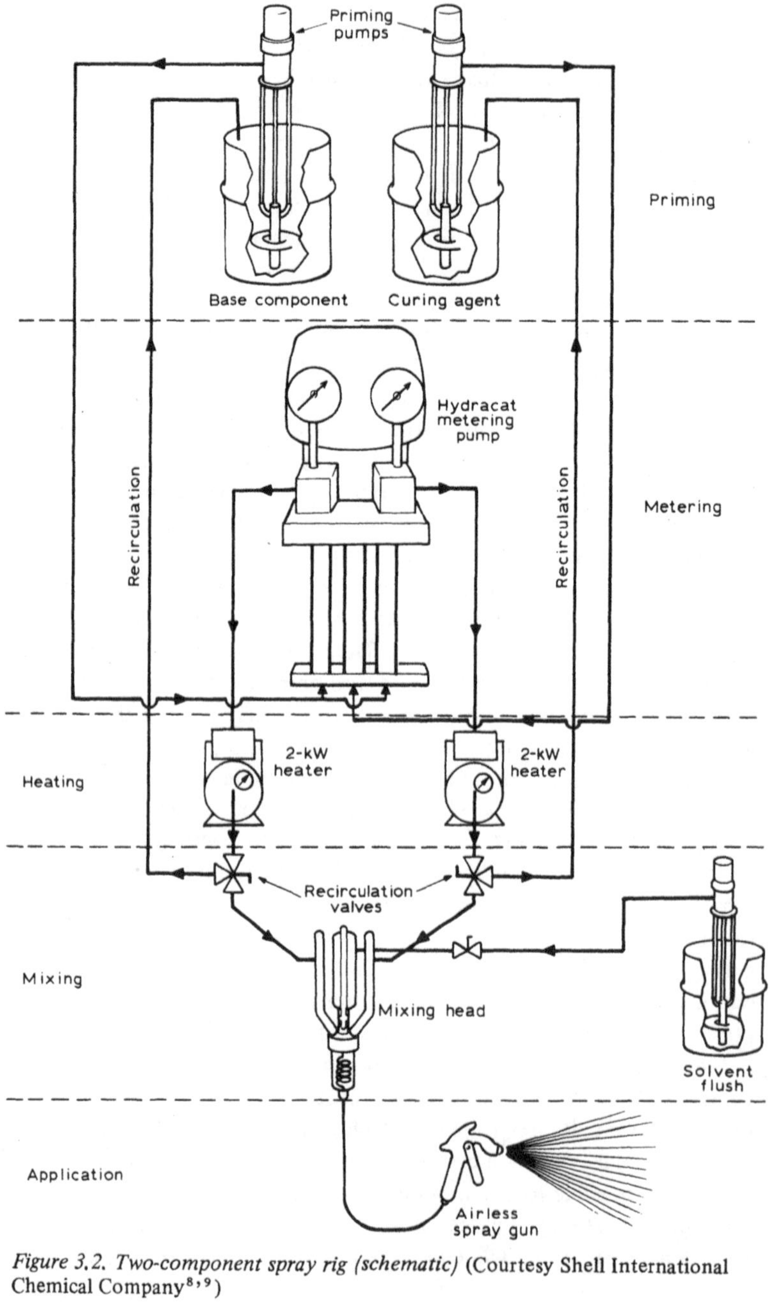

Figure 3.2. Two-component spray rig (schematic) (Courtesy Shell International Chemical Company[8,9])

range. Solventless paints are also more viscous than conventional paints and cannot be sprayed through normal equipment. Heating the mixed paint would not solve the difficulty, as it would also speed up the rate of cure of the paint and shorten the pot life still further. Adding diluents to the paint would reduce the viscosity but also reduce the performance of the coatings.

The problem therefore became one of applying a viscous reactive system, which will cure down to 0 °C, in a thick coating of up to 300 μm thickness as rapidly as possible. The solution to this problem came with the development of an airless, heated two-component spraying unit, able to pump, meter, heat, mix and spray a solventless epoxy system satisfactorily. A schematic diagram of such a unit is given in *Figure 3.2.* Two priming pumps feed the metering pumps with paint constituents under pressure and the positive displacement metering pump passes the constituents through two in-line coil heaters to the mixing chamber from which the mixture is fed to an airless spray gun. The equipment, although somewhat expensive, has a maximum output of about 3 litres of paint per minute[8,9]. There are now many manufacturers of equipment of this type who have developed spraying machines with the encouragement of epoxy manufacturers. Shell, who have been very active in the development of solventless epoxy paints, have published some interesting theoretical costings on the use of these spraying systems[9]. Thus, it is assumed that:

(a) capital cost of equipment is £4000;
(b) capital cost is depreciated over 3 years;
(c) the total maintenance costs over the 3-year period are £2000;
(d) the equipment is only used for 25 per cent of the available working time.

Therefore, the total depreciated cost over 3 years is £4000 + £2000, i.e. £6000, and the cost per annum is hence £2000.

$$\text{Total output of machine} = 3\text{–}5 \text{ kg/min} = 180\text{–}300 \text{ kg/h.}$$

At 25 per cent utilisation, the machine operates for

$$\frac{8 \times 5 \times 50}{4} = 500 \text{ h/year}$$

assuming a working year is 8 × 5 × 50 h. Therefore,

$$\text{Total amount sprayed} = 90\,000\text{–}150\,000 \text{ kg/year}$$

With a 3-year depreciation, therefore, the application cost is

2.2–1.3 p per kg of paint

i.e. 1.1–0.7 p/m^2 (300 μm thick)

This calculation takes no account of labour costs, but merely shows the low cost of the equipment per square metre of paint applied.

Using airless spraying equipment of this type, epoxy solventless paint can be formulated and applied with the following advantages over solvent-containing systems:

(1) Single coats can be applied in thicknesses of 3–400 μm, at up to 10 m^2/min, thus reducing the application time and giving economic coating of large areas.
(2) Accurate control of the application temperature and viscosity leads to outstanding consistency in film thickness and quality.
(3) The absence of solvents reduces (a) fire hazards, especially in confined areas such as ships' tanks, and (b) atmospheric pollution by solvents.
(4) Ventilation problems in confined spaces are also reduced.
(5) The paints will cure in reasonable times at low ambient temperatures, in the range 10–0 °C, depending on the system used.

EPOXY–COAL TAR SYSTEMS[10]

All amine-cured paints can be modified by the incorporation of up to 50–70 per cent of compatible coal tar pitch in the total binder content. The pitch acts as a diluent and also improves certain paint properties, and careful formulation can produce a very thick high-build system, which can give a dry film thickness of 4–10 mil (100–250 μm) in one application.

The advantages of epoxy coal tar paints are:

(1) Good alkali resistance and excellent water resistance.
(2) Very good flexibility, adhesion and hardness; the use of primers is often unnecessary.
(3) The ability to be applied in thick coats by brush or airless spray.
(4) Cheaper raw material costs compared with the unmodified two-pack epoxy systems.
(5) Excellent ability to withstand long-term exterior exposure to adverse environments.

These systems are now used widely for the protection of structural steelwork, especially when the steel is in contact with water.

ZINC-RICH EPOXY PRIMERS

A further important epoxy coating for protection against metallic corrosion is zinc-rich epoxy priming paint. Every day, thousands of man-hours and a great deal of money are wasted on preparing heavy-gauge steelwork before painting. The methods usually used – scraping, chipping and wire brushing – cannot remove deeply seated rust and scale, even when carried out conscientiously. Inevitably, the durability of paint applied to these surfaces must be less than that of paint applied to a well prepared surface.

The protection of heavy-gauge steelwork from corrosion by paint requires complete removal of all rust and scale, if the protection is to be successful and economic. Research work has shown that the durability of a paint system applied to well prepared (e.g. shot-blasted) steel may be as much as five times greater than the same paint system applied to poorly prepared steel. It is therefore important to clean steel by the most effective methods, which include sand and grit blasting. However, the grit blasting of steel after erection is expensive and frequently impracticable, and the preferred arrangement is to blast the steel before erection, at the steel mill or the fabricator's works, using automatic equipment. The primer should then be applied to the steel immediately after descaling.

The correct choice of the shop-applied primer is all important. The primer should:

(a) be easy to apply and dry within a few minutes in order to permit early handling;
(b) allow subsequent fabrication operations such as welding, drilling, cutting and rolling to proceed without difficulty;
(c) be resistant to bad handling;
(d) provide some measure of cathodic protection;
(e) give at least 6 months' protection on outdoor exposure;
(f) be suitable for overcoating with a wide range of paints, after an ageing period.

Epoxy zinc-rich primers that contain 92–95 per cent of zinc dust and 8–5 per cent of binder meet all of these requirements and have been in use for many years for the protection of structural steel. The use of the traditional zinc primer has been limited by its lack of resistance to salt water and, as the binder was usually a thermoplastic resin, loss of adhesion occurred and the primer was softened when overcoated with a paint that contained strong solvents. These difficulties have been overcome by the use of an amine-cured epoxy binder. In addition to the improved physical and chemical resistance, the epoxy system dries extremely quickly, which allows steel to be shot-blasted and primed continuously in ship yards and steel mills, so that the plate can be rolled shortly after priming. Steel primed in this way is thereby maintained in a clean and corrosion-free state during fabrication.

These epoxy primers and their use in many different situations are fully described elsewhere[11-15]. The epoxy–zinc systems became accepted at a time when, at least in shipbuilding, more operations were being moved under cover, and the primers allowed cost savings to be made and cleanliness of working to be achieved. This attracted competition from other primers, and now a number of different binders are used for zinc-rich primers.

SOME USES FOR EPOXY ANTI-CORROSIVE PAINTS

As noted earlier in the chapter, emphasis is now placed on the cost of providing a protective coating per unit area per year rather than the cost per gallon of paint, and the coating system to be used for any major undertaking must now be carefully designed, like any other component. This design requires a detailed analysis of surface preparation, application and coating requirements, and other factors that must be considered include the initial cost of the paint and the cost of its application together with subsequent repair and maintenance costs. It is through analyses of this type that epoxy paints have obtained their important place in the industrial maintenance and marine field over the last 20 years.

Examples of the use of epoxy anti-corrosive paints are therefore very numerous and widespread, although few are described in the current literature. In the general area of steel protection, epoxy primers and/or top coats have been used in chemical plant, oil refineries, marine installations, food factories, paper mills, gas holders and many other situations, including bridges, dams, lock gates and conveyor systems. The absence of literature reports is no doubt a sign of the general acceptance that these systems have gained in general maintenance. It is therefore the earlier literature of the late 1950s and early 1960s which contains most documented evidence of the economic advantages stemming from the use of epoxy maintenance paints. For example, in 1957 it was reported[16] that the Keystone Steel and Wire Company had been able to make substantial savings through the use of epoxy resins for plant maintenance. This company produces a wide range of wire products, such as nails, wire mesh and fencing. All the rods and wire are cleaned in vats of heated sulphuric or hydrochloric acid and the fumes from these vats readily attack the wooden structural members of the plant building, and no conventional coating was found that would last more than a few months in that environment, all of them peeling off and requiring replacement at regular intervals. Eventually, the wooden structural members were also replaced, and the new timber was coated with an epoxy–amine paint. This coating, at the time of the report, had lasted for a year without breakdown. The company, acting on this experience, then coated with epoxy paint all pieces of equipment such as transformers, heaters, ventilators and crane runways, in addition to structural steelwork.

A large U.S. soap and detergent company found that in their detergent plants, alkyd paints lasted about 1–2 months, chlorinated rubber paints lasted about 3 years and, at the time of the report, epoxy paint had lasted 4 years. The cost of applying either epoxy or alkyd coatings was 3.1 ¢/ft^2, whereas raw material costs were 1.5 and 0.9 ¢/ft^2, respectively. Although the alkyd therefore cost 4 ¢/ft^2 to apply compared with 4.6 ¢/ft^2 for the epoxy, the extra life of the epoxy more than offset this cost differential, especially when the cost of sand blasting and scaffold erection must be incurred on every repainting.

The same company made extensive savings by coating carbon steel tanks with epoxy resin rather than by using stainless-steel tanks. The comparative costs in \$/ft^2 were:

	Metal price	*Lining cost*	*Total cost*
Carbon steel with ¼ in thick epoxy coating	2.50	1.10	4.00
Stainless steel, ¼ in thick	15.00	–	15.00

In a similar way, five old detergent kettles, each 16 ft in diameter and 21 ft deep, were renovated with an epoxy lining at a cost of $3000 for each tank. Had the kettles been lined with stainless steel, the cost would have been $15 000 per kettle, and hence a total saving of $50 000 was realised for all five tanks.

Epoxy tank linings have been very widely used as substitutes for stainless-steel linings in order to save costs. Further, it has often been found possible to upgrade a concrete tank so that it may contain corrosive materials by providing it with an epoxy lining. Metal and concrete, marine, road, rail and static storage tanks are frequently coated internally in order to prevent corrosion and to maintain the purity of the contents, and epoxy paints are available that will withstand wine, beer, cider, molasses, alkalis, detergents and crude and refined oil products.

For small metal tanks, a stoved epoxy coating can be used, such as a phenolic-modified system, but tank lining is mainly carried out with one of the two-component ambient temperature curing systems which provide a high degree of protection. The final selection of the coating will, of course, depend on the resistance of the coating to the tank contents and also the results of extraction tests, especially when food products or aviation gasoline are to be contained in the tank. Thus, the coal tar-modified systems, while giving good protection when crude oil or sea water ballast is to be contained in the tank, could not be used for tanks intended to carry aviation fuels, when the 'gum' content of the fuel would be affected.

The exteriors of steel tanks can also be coated with epoxy paints, such as an epoxy ester primer plus alkyd finish, or epoxy ester primer and finishing coat. Alternatively, the two-component systems can be employed. Concrete tanks are often left unpainted externally, but in marine, smoke- or acid-polluted atmospheres, the two-component systems can be used for protection and maintaining appearance.

A recent application of epoxy paints to the second Mersey Tunnel at Liverpool (Plate 1), which was opened in mid-1971, typifies how epoxy coatings are selected in order to minimise costly repainting and maintenance[17]. The 2200 m long tunnel is constructed from concrete segments faced with steel plates, and these plates had to be protected from the corrosive effects of car fumes, high condensation and vigorous cleaning with detergents. The paint also had to be selected so as to avoid a costly maintenance programme. Accordingly, the plates were grit-blasted to white metal and a zinc-rich epoxy primer was applied, followed by a thick coating of a solventless amine-cured epoxy applied by hot spraying. In all, 50 000 litre of paint were used to coat the inner tunnel walls.

Two specialised and very important uses for epoxy protective paints are automotive primers and can linings. The motor vehicle industry represents one of the largest single markets for epoxy coatings, most of the resin being used as epoxy ester for primers or primer–surfacers. These primers protect the mild steel body of the vehicle against corrosion from water and salts and from the effects of flying stones. They were formerly applied by dipping or spraying, but now most vehicles are primed using the electrodeposition (ED) process, also known as electropainting. At present there are four types of resinous binder being used in base coats for vehicles:

(1) Epoxy esters.
(2) Phenolic–alkyd–butadiene combinations, sometimes with epoxy modification.
(3) Esters of drying oils and copolymers of allyl alcohol with styrene–resinous polyols.
(4) Maleinised drying oils.

The chief differences between the epoxy systems and the others are that the former possess better throwing power and are somewhat easier to control in the ED tank. The adhesion and corrosion resistance of all systems on phosphatised steel are very similar, but adhesion between primer and primer–surfacer coats and top coats is best with an epoxy base coat.

There are, of course, criticisms of the ED primers of all types: they are less resistant to stone chipping and offer inferior protection to corrosion spread compared with the spray-applied and dip primers. Paint manufacturers are therefore exploring a number of possibilities for overcoming these deficiencies, such as applying a thicker coat in the ED operation. Perhaps a more important drawback to ED and other solvent-containing primers in the long term is the problem of disposing of the solvent vapour or the exhausted aqueous systems from the ED baths. Vehicle manufacturers are therefore actively considering epoxy powder coatings as primer–surfacers for motor vehicles, although it is unlikely that they will be in use before 1975–1980. They would have the advantage over solvent- or aqueous-borne systems of having no disposal problem and, in addition, they could be applied as a thick film of perhaps 50 μm, with minimum wastage. Other benefits should also be attainable, such as improved inter-coat adhesion and chip resistance.

The packaging and container industry also consumes large amounts of epoxy-based paints for coating tinplate and aluminium cans, steel drums, collapsible metal tubes, bottle caps and other types of closures. Linings to cans are needed in order to preserve the characteristics of products such as foods and drinks, the flavours of which can be affected by direct contact with metals. Linings also protect the metal of the can from corrosion and permit the use of lightly tinned, less expensive tinplate. Adhesion, flexibility and chemical resistance are the most important properties of the can lining material. Good adhesion is needed to prevent the contents of the can from penetrating between the coating and the metal and the coating must withstand not only the mechanical forces exerted on it during the shaping and forming processes but also the temperature of soldering the side-seam of the can.

The various film-forming materials used for can lining include alkyd and oleoresinous varnishes, hydrocarbon resins, phenolics, vinyls, epoxy–amines, epoxy–ureas and epoxy–phenolics. The phenolics have high resistance to a wide range of chemicals but are brittle, have poor adhesion properties and are susceptible to scratching, and are therefore usually modified with an epoxy resin. Vinyls possess good flexibility but also have poor adhesion properties. However, vinyls have excellent chemical resistance and are therefore often used as a top coat over an epoxy or other base coat in applications that require the highest level of protection. The oleoresinous and alkyd coatings possess poor chemical resistance and are used for applications in which high performance is not required. Epoxy coatings are, in general, more flexible than the phenolics, have better heat resistance than the vinyls, and have the best adhesion to tinplate of all the can lining lacquers. Hydrocarbon resins can also form the basis of good quality can lining lacquers at a price competitive with that of epoxy systems and in the U.S.A. in particular, they began to be more popular than the epoxy systems. However, the introduction of the aluminium and then the tin-free can emphasised the disadvantage of the indifferent adhesion of hydrocarbon resins towards these materials, and the U.S. can manufacturers therefore reverted to using the epoxy–phenolic system.

Nowadays, very many different products are marketed in metal cans, ranging from foodstuffs, beer and soft drinks to paint, detergents and insecticides, etc. Each product presents its own particular set of problems to the can manufacturer, who must carefully select the can lining to be used for each specific case.

Interior coatings to drums are used to extend the range of products that can be packed in steel containers and the same coating materials that are used as can linings are also used for lining drums. However, the most widely used drum linings are the epoxy–phenolics, which have much better alkali resistance, adhesion and flexibility than unmodified phenolics and the vinyls. These physical properties are particularly important for drum linings because a brittle lining will chip if the drum is dropped or rolled on its hoops. Epoxy–phenolic-lined drums are used to contain a wide range of chemical products such as detergents, alkalis, dilute mineral acids, organic solvents, fatty acids, alcohols, oils and greases and food products.

Collapsible aluminium tubes are usually lined with an epoxy stoving ester lacquer. The most important property required of the coating in this application is the ability to withstand crushing of the tube, and the outstanding adhesion and flexibility of epoxy coatings are essential.

Clear epoxy lacquers are used on metal caps and closures of both the screw and crown types. Epoxy esters are used instead of alkyds, vinyls or oleoresinous coatings because they are tougher and more abrasion resistant.

REFERENCES

1. *Report of the Committee on Corrosion and Protection,* H.M.S.O., London (1971)
2. WOODYARD, D., *Engineering, Lond.* **211**, No. 5, 565 (1971)
3. HOLBROW, G. L., *Finan. Times,* (3rd. June, 1970)
4. *Coatings and Finishes Based on Araldite Resins,* CIBA-Geigy Ltd.
5. SCOGIN, E. R. and HOWARD, H. W., *Petrol. Refiner,* **37**, No. 1, 152 (1958)
6. *Epikote Resins in Maintenance Paints,* Shell International Chemical Company
7. CHALLIS, H., *Engineering, Lond.* **210**, No. 5460, 698 (1971)
8. *Resin,* Shell International Chemical Company, No. 1, p. 6
9. *Epikote Resin Solventless Coatings,* Shell International Chemical Company
10. *Epikote Resin Coal Tar Formulations,* Shell International Chemical Company
11. *Zinc Rich Primers Based on Amine Cured Epikote Resin Systems,* Shell International Chemical Company

12. *Corros. Prev. Control,* **9,** No. 11, Suppl. (1962)
13. JANES, T. K., *Paint Varn. Prod.,* **55**, 57 (1965)
14. WILSON, R. W. and ZONSVELD, J. J., Paper presented to N.E. Coast Section of Institution of Engineers and Shipbuilders, Newcastle-upon-Tyne, March, 1962.
15. *Epikote Resin Zinc Rich Primers for Structural Steel,* Shell International Chemical Company
16. McCABE, R., *Pl. Engng.,* **11**, No. 11, 106 (1957)
17. CORK, M., *Paint Technol.,* **35**, No. 9, 26 (1971)

PLATE 1
Use of epoxy solventless paint in the new River Mersey Road Tunnel (Courtesy of Shell)

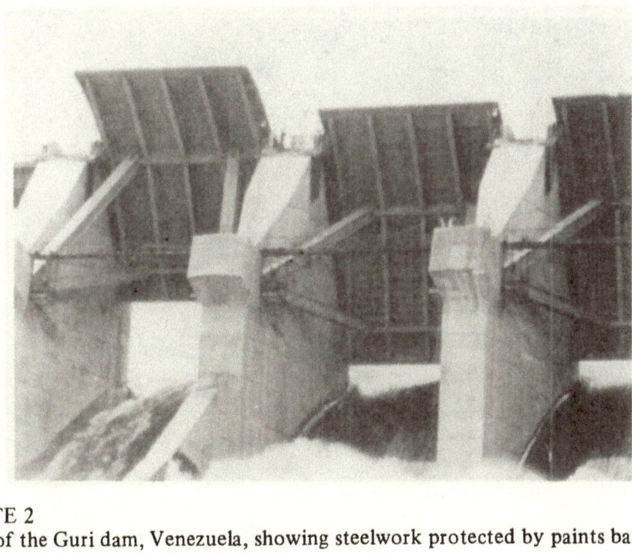

PLATE 2
Part of the Guri dam, Venezuela, showing steelwork protected by paints based on epoxy resin (Courtesy of Shell)

PLATE 3
Hull of tanker *Cinulia* which has been treated with an epoxy based paint (Courtesy of Shell)

4 Marine Coatings

More than ever before, shipping interests in the 'heavy' marine field are realising the economic importance of protecting vessels with a high-quality paint system properly applied on to a carefully prepared surface. In 1965, Klaren and Russell[1] estimated the cost of unchecked corrosion in an oil tanker fleet of 200 vessels to be £2.8 million per annum, and this figure has inevitably increased by at least 5–6 per cent per annum over the last 9 years. Factors which have brought the consideration of marine surface protection to the forefront have been:

(1) Maintenance painting of ships in dry-dock, especially the new giant oil tankers, must be kept to the absolute minimum in order to maximise the use of the large investment in the vessels.
(2) Painting at sea is now impractical because of the very small sizes of crews and shorter turn-round times.
(3) Dry-dock facilities are very expensive, especially for very large vessels.
(4) The use of cathodic protection, which has resulted in more stringent performance requirements for the paint film.

To place some of these factors in perspective, it has been stated[2] that the cost of paint alone for coating the exterior of a 170 000-ton supertanker would be approximately £140 000 and, in addition, if the tanks were lined, the total cost of painting would be about £300 000. Moreover, the overall costs of dry-docking for this size of tanker could be between £100 000 and £150 000 per week. Clearly, with costs of this order, it is essential to take the long view when approaching ships' maintenance. It has also been reported[1] that over a 5-year period, a tanker whose hull was scraped and repainted in dry-dock every year with conventional paints had to consume 20 per cent more fuel at the end of the period than at the beginning in order to maintain the same speed. This increase could only have been due to the increase in surface roughness of the hull due to corrosion, despite the annual repaint. If a large fleet operator has an annual fuel bill of about £10 million per annum, even a slight saving in this cost would be significant.

The paints available for marine protection include those based on alkyds, vinyls, chlorinated rubber and polyurethane resins, as well as the epoxy systems. However, the epoxy systems have a unique set of properties that make them superior to the other coatings: they have better alkali resistance than the alkyds and better solvent resistance than the vinyls and chlorinated rubber paints, and they have outstanding adhesion to most substrates, and can be applied in very thick single coats. Their limitations are poor resistance to ultra-violet light and slow rates of cure at temperatures below 10 °C.

PRIMERS

For priming paints on ships' plates, zinc-rich epoxy primers with high or low zinc loading are widely used. Other types of primer available include bodied wash vinyls and inert pigmented epoxy two-pack primers. The vinyl wash primers are, to some extent, water sensitive and the inert pigmented epoxy primers have no electrochemical protective effect but provide a physical barrier against corrosion. The zinc-rich epoxy primers are undoubtedly the primers to be preferred, giving better long-term protection to hulls when properly overcoated, than any other primer available. These primers dry quickly, have excellent adhesion and welding properties and chemical resistance and are ideal for most types of paint systems.

Occasionally, difficulties have been experienced with zinc-rich primers, for example when overcoating a weathered primer surface. It is essential, in this situation, to treat the primer surface as if it were an old paint film, and to clean and wash the surface carefully before overpainting it. The first coat over the primer should also thoroughly wet and seal the surface to as to ensure good adhesion, and to avoid blistering due to loss of cohesion in the primer because of the low binder content.

Other difficulties, such as blistering and detachment, reduced cutting speeds for plates, porosity of welds, and fumes during cutting and welding, have also occurred. Generally, these have been overcome by ensuring that the total thickness of the primer film does not exceed the maximum recommended by the paint manufacturers, which is usually 30 μm.

Table 4.1

A COMPARISON OF ZINC-RICH EPOXY AND VINYL WASH PRIMERS (AFTER SHELL INTERNATIONAL CHEMICAL COMPANY[2])

Paint systems		*Film thickness, µm*	*Abrasion resistance, % of paint surface abraded*		*Paint performance after 4000 h rotation in artificial sea water**	*Resistance to cathodic protection after 2000 h testing in artificial sea water**	
Primer	*Topcoat*		*After 1 h testing*	*After 2 h testing*		*Normal protection, −850 mV*	*Over-protection, −1000 mV*
Epikote† zinc rich (one coat)	Epikote coal tar (two coats)	260	6	29	9	9	8
Vinyl wash primer (one coat)	Epikote coal tar (two coats)	260	33	54	7	7	2
—	Chlorinated rubber (five coats)	200	53	72	5	5	1
—	Aluminium–bitumen (five coats)	250	43	64	3	6	1
Epikote zinc rich (one coat)	Wood oil–phenolic (three coats)	160	56	67	6	3	1
—	Acid-free tar (eight coats)	200	77	85	1	2	1

*Key: 10 = excellent; 0 = very poor.
†Shell Chemicals epoxy resin.

In some situations of short production schedules, the reduced period for which the primer must provide protection before overcoating allows a zinc-rich epoxy primer to be used with a reduced (70 per cent) loading of zinc. These coatings do not give the same degree of protection as those with high zinc loadings, but they are cheaper and easier to overcoat after weathering. An indication of the performance of a zinc-rich epoxy primer compared with a vinyl wash primer is given in *Table 4.1*; the zinc-rich epoxy primer with an epoxy coal tar top coat was clearly superior to the other paints considered.

Zinc-rich epoxy primers are now used for ships' primers in many different countries. Thus, in the U.S.A., shop or prefabrication primers in the marine field are most often zinc-rich epoxy systems[3] and the same situation occurs in Japan. Sakami[4] and Goto[5] described the use of these primers in Japan and stated that 1680 tons of primer, covering 8 370 000 m^2 of steel have been applied to 116 vessels in the period 1965–1969 without any difficulty. The technique for applying the primers under protective cover in European shipyards has been fully described by Klaren[6, 7]. The steel is first shot-blasted and coated by automatic spraying with 12–15 μm of zinc epoxy primer. The steel is then fabricated into units which are cleaned and sprayed with a second 12–15 μm coat of the primer using airless spray equipment. The prefabricated units are painted with the top coat system, leaving a 10 cm unpainted strip where the welding seams will occur. After welding into the completed structure and ultimately the complete vessel, the seams are cleaned and coated with the top coat paint system.

HULL COATINGS

The effect of corrosion on a ship's hull and the increase in power (and therefore cost) required to maintain the same speed as the hull becomes more corroded has already been mentioned. It is therefore important to provide the hull with maximum protection against corrosion. The most successful hull paints for large vessels in recent years have undoubtedly been based on epoxy–coal tar combinations. The reasons for this are probably:

(1) The excellent water resistance of the system.

(2) The high abrasion resistance and smooth surface the paint provides.
(3) Excellent adhesion in cathodic protection situations, especially where over-voltages occur locally on the hull.
(4) The ability of the system to be applied in thick coats using airless spraying, and the absence of curing difficulties.

The epoxy–coal tar hull paints form a hard, smooth and largely impermeable barrier film on the hull, which is able to protect the hull for long periods without maintenance. For the underwater part of the hull, the coating most widely accepted is two coats of the epoxy–coal tar system, usually applied by airless spray to give a total thickness of about 250–300 μm. A period of 24 h is usually left between coats. As a final finish, one coat of chlorinated rubber or vinyl anti-fouling paint is then applied. Shell Chemical Company estimate that the saving in fuel costs over a 4-year period for a tanker coated in this way compared with a tanker coated with the earlier conventional paint systems is 5–7½ per cent[2]. *Table 4.2* gives fuller comparative costings of coating the hull of a tanker with the two paint systems.

On the boot-tops and topsides, where the hull is above the water line, appearance also becomes important as well as protection when choosing a paint. It is usual to overcoat the epoxy–coal tar paints with an ordinary epoxy–polyamide paint which does not contain coal tar. Alternatively, a solventless epoxy coating can be used, which is able to provide the desired 250 μm thick coating in fewer coats. The method of applying such a high-build coating is important, especially if painting is to be carried out at low temperatures in an exposed situation. Hot airless spraying of the paint is the preferred method, as it enables a solvent-free paint which is frequently viscous to be applied at low temperatures and ensures that cure takes place in a reasonable time because of the preheating of the paint components in the machine. This applicational technique has been described earlier (p. 43).

DECKS AND SUPERSTRUCTURES

The superstructure of ships is subject to corrosion from salt spray, all types of weather conditions and also possibly cargo spillage. In addition, decks are continually abraded by foot traffic and cargoes.

Table 4.2

COMPARATIVE COSTS OF HULL COATING: EPIKOTE*–COAL TAR *versus* ALUMINIUM–BITUMEN (AFTER SHELL INTERNATIONAL CHEMICAL COMPANY[2])

33 000 TON TANKER WITH A HULL AREA OF 7700 m^2: BASED ON NETHERLANDS REGISTERED VESSEL IN SERVICE 1965–69

Fuel costs				
Average fuel cost per day		£700		
Days in service per year		250		
Fuel cost over a 4-year period		£700 000		
Coating costs	*Aluminium–Bitumen*		*Epikote–coal tar*	
Preparation of steel	(Brushing)	£1950	(Blasting)	£9 600
Application of paint		£1950		
Paint		£1800		£3 850
Total		£5700		£13 450
Maintenance costs				
Cost per year	£1400 (normally 25–30% of initial painting cost)		£600 (normally less than 5% of initial painting cost)	
Cost per 3 years		£4200		£1800
Total costs over 4 years		**£9900**		**£15 250**
Extra cost of using Epikote–coal tar			£5350, approx. 0.75% of fuel costs over a 4-year period†	

*Shell Chemicals epoxy resin.

† In practice, the usual saving on fuel over a 4-year period using Epikote–coal tar is 5–7½ per cent.

The coatings used on these surfaces must therefore withstand these attacks and also be easy to apply to the different materials of which the superstructure is composed, e.g. wood, steel or aluminium. They must also give a good surface appearance.

Epoxy resin-based paints have proved themselves able to withstand the most exacting conditions, but a number of other paint systems are also used, such as polyurethanes. These latter paints exhibit good weather resistance and gloss retention, and can be used as a top coat over an epoxy primer and epoxy intermediate or undercoat. Alkyd resin-based top coats are also still used on some ships. On decks in particular, chlorinated rubber paints are also suitable for top coats. Epoxy-based non-skid surfacing materials have been used on many decks, the system being very similar to those described in the chapter dealing with flooring (Chapter 11). Such a surfacing could be used in certain situations in ships' interiors as well as in the more exposed situations.

SHIPS' TANK LININGS

Cargo tanks and holds can present a serious corrosion problem. Tanks may contain a wide variety of materials, some of which are highly corrosive, and tanks are often repeatedly cleaned with hot or cold sea water, sometimes with added detergent. If tanks are not protected, extra strength must be built into the tanks so as to compensate for corrosion of the steel, which increases costs and decreases the initial carrying capacity of the vessel. Moreover, when a tank contains white oil products or chemicals, it is essential that the product should not become contaminated with rust or other substances produced by corrosion of the unlined tank. A coated tank is also much quicker to clean when a rapid changeover from one type of cargo to another is necessary.

The corrosion of tanks can be combated by several well known methods:

(1) Cathodic protection.
(2) The use of water- or oil-soluble inhibitors.
(3) The application of coatings.

Of these three methods, coating with epoxy-based paint is to be preferred, despite its initial higher cost, for the following reasons[8]:

(1) The paint can provide 100 per cent protection against corrosion and wastage of steel is prevented. Coated tanks are therefore expected to last for 20–25 years without renewal of the steel, whereas other forms of protection will provide only 50–65 per cent protection and renewal of the steel with therefore be required during the working life of the tanker.
(2) Coatings are effective in both cargo and ballast space. Cathodic protection is only really effective under ballast conditions.
(3) There is less chance of product contamination from a coated tank than from a tank protected by other methods.
(4) The time required in order to clean a coated tank is usually shorter than for an uncoated tank.

Comparative economic data have also been published[8] on the cost of cathodic protection and the cost of painting with an epoxy paint, and showed a clear advantage for the epoxy coating as far as a white oil tanker is concerned.

Typical treatments of the various tanks of an oil tanker have been summarised[2] as follows.

Crude oil tanks are not normally completely coated. One method is to coat the bottom and deck-head plus about 1 m up and down the walls with 250 μm of an epoxy–coal tar paint. Crude oil–ballast tanks have the coating carried down further from the deck-head to about half the tank depth of 12 m. Clean-water ballast tanks are coated in a similar fashion.

In *product tanks*, the wide variety of products that may be carried, ranging from oil products such as gasoline to chemical solvents, vegetable oils, latices etc., makes it very difficult to be specific about the type of paint that should be used. Not only should the tank be protected from corrosion and the tank lining be impervious to any attack from the product being carried, but there should also be no possibility of contaminating the product with soluble materials either leached from the tank lining or formed by breakdown of the lining. In general, there is no universal tank liner, but the combination of a zinc-rich epoxy priming coat followed by a high-build or solventless epoxy system provides the best combination of properties. Inorganic zinc silicates are also used for tank linings, especially in the U.S.A., but they suffer from some disadvantages compared with epoxy systems[7,8]. For example, zinc silicate coatings require a much higher standard of surface preparation than epoxy coatings, dry slower, have poorer alkali resistance and cost more to apply, although silicates should be used for tanks containing certain strong solvents such as benzene or acetone.

SHIPS' INTERIORS

Boiler rooms, galleys and engine rooms are all internal locations in a ship where the conditions justify the use of an epoxy paint to protect against corrosion and spillage of fuel and food. The ability of epoxy coatings to withstand hard wear and repeated cleaning also leads them to be used in staterooms, restaurants, bedrooms and bathrooms.

In summary, it can be fairly claimed that epoxy resin paints are to be found on every major new ship constructed, and many ships now sailing, especially oil and chemical tankers, are virtually exclusively painted on the exterior with epoxy-based paints. Thus the largest tanker in the world at its time of launch in Sweden in 1969, the *John A. McCone*, 212 000 tons dwt, was covered with 130 000 m^2 (1.4 million ft^2) of coatings that were almost 100 per cent epoxy based[9].

REFERENCES

1. KLAREN, H. and RUSSELL, G. C. C., *Shipp. Wld*, **154**, No. 3747, 643 (1965)
2. *Modern Marine Painting with Epikote Resins*, Shell International Chemical Company
3. BLOODGOOD, D. T., *Paint Varn. Prod.*, **59**, No. 12, 27 (1969)
4. SAKAMI, T., *Anti-Corros. Meth. Mater.*, **16**, No. 12, 13 (1969)
5. GOTO, M., *Ind. Finish., Lond.*, **22**, No. 267, 12 (1970)
6. KLAREN, C. H. J., *Surface Coatings*, **4**, No. 5, 150 (1968)
7. KLAREN, C. H. J., *Surface Coatings*, **4**, No. 7, 218 (1968)
8. *Surface Coatings*, **1**, No. 2, 49 (1965)
9. *Anti-Corros. Meth. Mater.*, **16**, No. 12, 20 (1969)

5
Coatings for Steel Pipes

It has been estimated that in the period 1961–84 the world-wide consumption of energy will more than double, the bulk of this increase being provided by natural gas and oil as primary sources. With this increase in consumption of gas and oil, there will be a corresponding increase in the pipelines laid in order to transport these products. The accompanying general increase in industrial activity will also result in more pipelines being laid in order to transport water, chemical products, sewage and chemical effluent.

There are at present well over 250 000 miles of oil pipelines and 600 000 miles of natural gas pipelines in use, and these mileages are growing rapidly each year. It has been reported[1] that 50 000 miles of new oil and gas pipelines are being laid annually, at an annual cost of £1000 million. In addition, many miles of effluent, chemical and water pipelines and internal factory pipelines are being installed. Moreover, in recent years there has been a great increase in the amount of submarine pipe used for offshore installations. In 1968, it was estimated that 2000 miles of this type of pipe were in use, and the amount has grown rapidly since then. All of this pipe requires external coating and much also needs internal lining.

Looking into the future, the transport of solids such as grain, sulphur, steel, and aluminium by pipeline is likely to occur, the solid being encapsulated in a container which is suspended and moved by a carrier fluid such as water. Pipes carrying these materials will certainly require very tough, abrasion-resistant linings.

The pipeline field therefore presents a very attractive market for epoxy coatings, especially as the pipeline industry, more than ever before, is aware of the economic necessity of ensuring long-term, trouble-free service from pipes and is turning to epoxy resin paints to provide high-quality, durable protective coatings.

INTERNAL PIPE LINING

Much of the steel pipe currently in service is unlined. However, this situation is changing rapidly and more pipes are being coated with epoxy-based paints for the following reasons:

(1) To reduce internal corrosion
(2) To reduce external corrosion
(3) To reduce internal erosion

} and hence reduce maintenance and replacement costs. In 1967, the U.S. National Bureau of Standards estimated that the extra pumping costs due to rust clogging the interior of water lines was £20 million per year.

(4) To minimise product contamination.
(5) To reduce the surface–liquid coefficient of friction and hence increase the throughput for a given pipe diameter and power input.

The internal coating of steel pipes not only protects against corrosion and abrasion and hence reduces maintenance and replacement costs, but also reduces the cost of scrubbers, strainers and general pipeline cleaning. These advantages are additional to the increased throughout (5–8 per cent) and decrease in product contamination that can occur. The amine and polyamide-cured two-pack epoxy system and, more recently, solventless epoxy systems are widely used for internal linings for pipes that carry natural and wet gas, sour crudes, salt and fresh water, petroleum products and chemicals, and many specifications are based on epoxy coating systems. The coatings are applied either *in situ* or to each section of pipe before jointing. The *in situ* method, which was fully described by Bassett[2], has been used in the U.S.A. since 1947–48. It consists of first cleaning the pipe by passing brushes or scrapers through it, followed by cleaning with acid using special techniques. This is followed by light phosphating, washing with water in order to remove residual free acid, and then drying with solvent and finally hot air drying in order to remove the solvent. The epoxy paint is then applied between two special rubber plugs that are propelled through the pipe by compressed air. This method has been used successfully to coat continuous pipe lengths of up to 15–20 miles. Spray techniques have also been used to coat existing pipelines.

Probably the most extensive use of epoxy-lined pipe has occurred in the many gas pipeline networks around the world. In The Netherlands alone, it has been estimated that there are well over 1000 km of such pipes currently in service[3]. In these cases, the lining is not intended primarily for corrosion protection, but rather for increasing the throughput of gas. The thickness of the paint film is therefore only about 50 μm. The first major test of the effectiveness of the coating on throughput of gas was made at Refugio, Texas, in 1958 on a 12-mile length of 24-in pipe. An overall increase in gas flow efficiency of 12 per cent was obtained when coated pipes were used instead of uncoated pipes and of this increase, 7–8 per cent was attributed to cleaning the pipe and 4–5 per cent to the coating providing a smooth surface. Many subsequent trials have confirmed these findings. Thus Kut[4] reported Johnson's claim that a difference in flow efficiency of 11 per cent can be expected in a new coated pipe compared with an uncoated pipe exposed to the atmosphere for 6 months.

A further advantage of the internal coating of old lines is that gas leaks in the pipe are sealed by the paint. Cummings[5] quoted a 93 per cent reduction in leaks in a 2-mile section of an old pipe after coating the pipe internally.

Examination of epoxy-lined gas pipes that have been in service for up to 13 years has shown them to be in very good condition. A recent example of epoxy-lined pipes for transporting natural gas is the use by the Gas Council to construct a nationwide pipeline transmission system of 2500 miles in the U.K. to convey natural gas from the North Sea to various distribution points throughout the country. Some of this network is already in position, having been constructed during the 1960s, and work on the major remaining part of the pipeline is in progress. The coating with an epoxy paint of a 110-mile section of one of the main high-pressure transmission pipelines of the grid connecting Bacton to Stevenage has been fully described[6] and typifies the application of epoxy linings to large steel gas pipes. The pipes have an internal diameter of 36 in and are in 40-ft sections. They are first shot-blasted externally, followed by the internal coating process. A sprayhead and cleaning unit are supported at one end on a trolley running on rails, the forward movement of the trolley thrusts the unit into the revolving pipe and a rotary wire brush, turning in the opposite direction to the rotation of the pipe, thoroughly cleans the internal walls of millscale and other contaminants. The wire brush operates during the forward stroke of the unit, and the epoxy coating is sprayed during the backward stroke. The central member of the apparatus is a suction pipe, which is used to draw off the millscale from the internal surface of the pipe resulting from the action of the wire brush. Simultaneously with each motion of the unit inside the pipe, a nozzle supported on a rail on an overhead gantry sprays a coat of chlorinated rubber primer paint on to the outside of the pipe. This primer is subsequently covered with alternate layers of glass-fibre tissue and coal-tar pitch, which is applied hot. The internal epoxy coating used is a two-pack amine-cured system pigmented with light-fast and rust-inhibiting agents. It also contains additives to promote high build and quick setting properties, and is fully cured in 7 days.

In addition to gas pipes, epoxy coatings are also widely used as internal coatings for pipes carrying many different corrosive materials. Some examples of the successful use of epoxy lining, quoted by Kut[4], are given below.

Crude oil lines

An old submarine pipe required solvent 'deplugging' of paraffin deposits every 7 h, and the epoxy coating stopped the plug being formed. In another case, 100 miles of old gathering lines, having been coated with an epoxy resin, were in a satisfactory condition after 10 years. Further, 5 miles of new welded 4-in pipe were free from leaks and paraffin deposits after 4 years; prior to coating, the lines had been scrapped every 2 years and the paraffin removed at weekly intervals.

Potable water lines

A 3½-mile length of welded 4-in pipe coated in 1952 showed no deterioration after 8 years.

Salt water lines

A tanker-deballasting 24-in line, 2 miles long, was found to be satisfactory after 9 years' service. Epoxy paint can also rehabilitate leaking pipes. Thus, a 6-in line carrying aerated 10 per cent salt water plus hydrogen sulphide and carbon dioxide contaminants at pH 5.8, 65 °C and 4–6 atm was found to be leaking after 10 months' use, and heavy corrosion was also found to have occurred, causing a groove 1–1½-in wide and ⅛–¼-in deep along the bottom of the pipe. A coating of epoxy paint stopped the leakage of water and no further trouble was experienced.

Effluent and outfall lines

Many of these pipes are constructed of concrete or asbestos cement and, when lined, are coated with an epoxy–coal tar paint to a thickness of about 10–15 mil (250–373 μm).

The lining of sewage pipes is very important in warm climates, where the corrosive action of the acidic constituents formed may give rise to pitting. A considerable amount of sewage pipe in the south of the U.S.A. is lined with the epoxy–coal tar system, and many European firms are also now using pipe linings of this type. Thus, the 2½ miles of sewage outfall pipe at Baglan Bay in South Wales consists of 49-in diameter, ½-in thick spirally welded steel pipe coated internally with 0.02 in and externally with 0.01 in of an epoxy–coal tar system[7].

Power stations and hydroelectric schemes are large users of lined steel pipe. For example, the Ffestiniog, Wales, pumped storage scheme contains 5 miles of steel pipe, all of which is coated with a total of 50 tons of epoxy–coal tar paint[8].

Steel pipes for use under aggressive conditions can also be lined with an epoxy powder coating[9]. The largest steel pipe producer in Belgium, Usines à Tubes de la Meuse, carried out an extensive screening programme to decide upon an internal coating to be applied to pipe on the production line. Powders were considered in order to obtain high build in a single coat without having any pot life limitations. Various polymer-based powders were examined, for example polyamides, chlorinated polyethers and epoxy resins, but finally it was decided to use epoxy-based powders. One of the longest and most modern pipe lining installations was therefore set up at Flémalle, which can line pipes of diameter 10 cm–1.3 m. The coating procedure consists of grit-blasting and vacuum cleaning the pipe, followed by pre-heating the pipe with gas burners, the pipes being rotated on a roller bench. When the metal has reached the desired temperature, an electrostatic spray-gun mounted on a long rigid powder supply tube is inserted down the full length of the pipe and switched on. The spray-gun is moved backwards through the revolving pipe and the speed of traverse and rate of powder flow are adjusted so as to obtain the desired coating thickness without wastage. When the lining operation is completed, the pipe is transported to an oven and the coating cured by the appropriate curing schedule. Large pipes over 1 m in diameter are sprayed with a powder that is fully cured by the residual heat in the pipe, thus removing the need for post-curing in an oven.

EXTERNAL COATINGS

External epoxy pipe coatings are only now coming into widespread use. For economic reasons, most steel pipelines used in conventional service were hitherto coated externally with coal tar or bitumen. Now the epoxy–coal tar system is being used, reinforced with glass-fibre or other similar inorganic reinforcement, and many pipes in Australia, for example, are protected externally in this way[10].

Kut[4] reported the satisfactory coating with 12 mil of a special anti-abrasion epoxy coating of a pipe that was forced hydraulically through the bed of the Haarlem–Amsterdam highway as part of the pipeline crossing in the Veerpolder. Tests have also been carried out by The Hague authorities on epoxy coatings for sewer outfall lines.

Powder coatings have also been used for external pipe protection. In fact, it is intended to coat with an epoxy powder part of what is probably the largest projected oil pipeline system in the world, namely the 400 miles of underground 48-in (1.22-m) diameter oil pipeline from the Prudhoe Bay field in North Alaska to the warm-water port of Valdec near Anchorage[11]. This line if built will be capable of carrying 2 million barrels of crude oil per day. The pipes will be coated with a 10-mil thick external coating on site, the pipes being heated and coated, and the powder being cured with the residual heat in the pipe.

The use of epoxy powders for external pipe coating has become particularly important in the U.S.A., especially for pipes of smaller diameter. By the late 1960s, over 1300 miles (2090 km) of small-diameter pipe (1–8⅝ in) had been coated by one contractor alone, and the costs in *Table 5.1* show the advantage in using the epoxy powder[3]. These data compare the applied cost of an epoxy powder with a 3/32-in (2.5-mm) coal tar coating reinforced with glass-fibre.

Table 5.1

COMPARATIVE PIPE COATING COSTS (US DOLLARS PER 100 ft LENGTH) (AFTER SHELL INTERNATIONAL CHEMICAL COMPANY[3])

Factory applied : car load lots.

Pipe diameter, in	*Epoxy powder coating*	*Coal tar enamel*
1	9.15	10.02
1½	11.25	13.22
3½	17.46	19.89
4½	23.01	25.64
6⅝	36.36	35.26
8⅝	50.45	45.07

The epoxy powder coating also provides better corrosion resistance, little 'burn back' on welding and a lighter weight of pipe. Most important is the superior resistance of epoxy systems to cathodic disbanding, which is a considerable problem with cathodically protected pipes laid underground. Recent experience has shown that care must be taken to ensure that the surface of the pipe is clean and free from contaminants before the epoxy powder is applied. It also appears that the best results are obtained if the epoxy powders are applied over an epoxy primer.

Kut[1] has reported a similar assessment by Neal in 1968 on the use of epoxy powders for the external coating of steel pipe. Neal found that a complete epoxy system (powder coating plus solvent-free epoxy weld repair) offered a price advantage over a conventional external coating system at pipe sizes of 6⅝ in (16.8 cm) diameter and below. At 8⅝ in diameter the costs were the same, but at 6⅝ in (16.8 cm) diameter the epoxy system was 5–8 per cent cheaper and at 4 in (10.2 cm) diameter the epoxy system was 20–23 per cent cheaper. This assessment was carried out in 1968, and since then epoxy powder prices have certainly fallen as the volume produced has increased. The economics are therefore probably even more favourable for the use of epoxy powders than ever before. On very large diameter pipes a similar trend towards epoxy systems becoming competitive on the basis of price as well as performance has also occurred.

REFERENCES

1. KUT, S., *Pipes Pipelines Int.* **17**, No. 2, 8 (1972)
2. BASSETT, A. R., *Pipes, Pipelines Convention, London,* (1962) Paper 2g

3. *Epikote Resin Coatings for Pipelines,* Shell International Chemical Company
4. KUT, S., *Pipes, Pipelines Int.* **12,** No. 7, 21 (1967); **14,** No. 2, 31 (1969); **14,** No. 4, 40 (1969); **14,** No. 6, 41 (1969); **14,** No. 7, 45 (1969)
5. CUMMINGS, H., *Oil Gas J.,* **72,** No. 54, 243 (1956)
6. *Araldite in North Sea Gas Pipelines,* CIBA-Geigy (1968)
7. *Ind. Finish.,* **22,** No. 261, 46 (1970)
8. CHALLIS, H., *Engineering, Lond.,* **210,** No. 5460, 698 (1971)
9. *Resin,* Shell International Chemical Company, No. 2, p.13
10. *Aust. Chem. Engng.,* **10,** No. 9, 33 (1969)
11. *Product Finish.,* **24,** No. 8, 18 (1971)

6

Epoxy Powder Coatings

'The Alternative to Chromium Plating is Coating with Epoxy Powder'

'With Epoxy Powder, Who Needs Nickel at £1500 per Ton?'

(Advertisements in trade journals)

Powders represent one of the fastest growing sectors of the paint market in Europe, and probably in the world, and epoxy powders now form a large part of the total amount of powders consumed. The technique of applying a 100 per cent solids powder coating to a surface is not new, having been known and practised for some 30 years using mainly powders based on thermoplastic materials such as PVC, polyethylene and nylon. Why, then, has this fairly sudden increase in the use of powders occurred?

The main reason can probably be found in the research work carried out in the early 1960s by a few European epoxy resin manufacturers, notably Shell. This work led to the development of powder coatings based on epoxy resins which were technically superior both to thermoplastic resin systems and also to conventional solvent-borne paint systems, and which could be applied by the usual fluidised bed or flame spraying techniques. A further important influence on the growth in powder consumption has been the development of electrostatic techniques for applying the powders. But no doubt even more important for the long-term future of powders are their economic advantages and the ever-growing concern over the pollution caused by exhausting solvent vapours into the atmosphere. A 100 per cent solids paint system eliminates this problem completely, and legislation in many countries will no doubt gradually force manufacturers and paint suppliers to use powders for most industrial painting. Moreover, not only are powders, notably epoxy-based, actively replacing solvent-borne stoving paints, but they are also competing with vitreous enamel as a hard-wearing surface finish for metal.

The advantages of powder coatings, in general, over stoving paints and vitreous enamel have been discussed very fully by Stephens[1], and can be summarised as follows:

(*a*) *Technical:*

(1) Powders are a 100 per cent solids system. The absence of solvents removes the need for a flash-off time before the coating is stoved.
(2) Powders can be applied in thick films in one application, which is equivalent to two or more coats of a liquid paint.
(3) Nearly 100 per cent of the powder can be used by recycling the overspray, whereas with solvent-borne paints up to 60 per cent of the paint can be lost during spraying.
(4) Sags and runs are eliminated from the paint film and edges and corners are better covered.
(5) Compared with stoving paints, hard, smooth films with better chemical resistance, dielectric strength and abrasion resistance are obtained.
(6) Dust-free films are more readily obtained.

(*b*) *Safety and health:*

(1) Because of the absence of solvents, fire and explosion hazards are reduced during both powder manufacture and applications.
(2) Solvent removal and solvent pollution of the atmosphere by enting solvents into the air are eliminated by the use of powders.
(3) Working conditions for the operators are improved by the absence of toxic vapours and objectionable odours.

Similar comparisons can be made between powders and vitreous enamel. Apart from the much lower application temperatures (130–160 °C compared with 750 °C) and the consequent restriction of vitreous enamel to articles that can withstand this temperature and undergo no dimensional change, powders can have vastly superior flexibility and chip and impact resistance. The very high reject rates of enamelled articles do not occur with powder-coated items and the several preparatory coatings required before enamelling are also not necessary. Thinner films are possible with powders compared with the 15 mil of vitreous enamel needed and powders can more easily coat corners that have a small radius.

Sufficient has perhaps been given to convince the reader that powder coatings, as a class, do have advantages over solvent-borne stoving paints and vitreous enamel, and an economic analysis that bears this out is given later in the chapter. But why have epoxy powders assumed such a dominant position in the field, and how do they compare with the other powders?

Thermoplastic powders, although widely used for many years, have a number of limitations to their use:

(1) Poor adhesion to substrates. Most thermoplastic powders tend to form a skin or sleeve round the object they are coating rather than a film that is integral with the metal surface. Once this skin has been broken, the coating rapidly deteriorates. Very large film thicknesses have therefore been used in order to achieve some degree of stiffness in the coating which will mask the poor adhesion.
(2) A tendency to soften at elevated temperatures.
(3) Cracking under thermal cycling conditions.
(4) High application temperatures, between 200 and 480 °C.

Powders based on thermosets such as epoxy, acrylic and polyester resins do not, in general, have these limitations, and also offer a number of other advantages. This is particularly true of epoxy-based powders, which represent over 90 per cent of all the thermoset-based powders currently sold. Epoxy powders have higher gloss, superior adhesion, greater hardness and scratch resistance, and a wider range of chemical and solvent resistance than thermoplastic powders, and their curing-oven temperatures are lower and they have better resistance to thermal cycling.

The limitations to the use of epoxy are chiefly concerned with their external use. It is well known that all DPP-based epoxy coatings will chalk on exposure to ultra-violet light and suffer a loss of gloss. These powders are therefore used on items where chalking can be tolerated, or where the outdoor exposure is intermittent. This does, however, have important implications for the future penetration by epoxy resins in the large volume outlets in the automotive and appliance industries. Acrylic powders have now been developed that possess good gloss retention on external exposure and colour retention at elevated temperatures and no doubt they will provide strong competition for epoxy powders in exterior applications and for coating items such as motor vehicles, appliances, radiators and oven and boiler casings. Curing temperatures for the acrylic powders are, however, higher than for epoxy powders, typically being 20 min at 200 °C. Other thermoset powders that are available include those based on polyesters and alkyd–melamine mixtures. The polyesters, in general, have gloss and gloss retention properties that are intermediate between those of the epoxy and acrylic systems, but do not match epoxy systems in their physical and chemical resistance properties. Similarly, the alkyd powders possess good yellowing and weather resistance characteristics, but need curing at 180–200 °C for 30 min and do not have a high level of chemical and solvent resistance.

A simple comparison of the properties of powders based on four different resins is given in *Table 6.1.*

Table 6.1

PROPERTIES OF POWDERS BASED ON DIFFERENT RESINS

Key: E = excellent; G = good; P = poor.

	Resin			
Property	*Acrylic*	*Epoxy*	*Polyester*	*Alkyd*
Flexibility	G	E	G	P
Adhesion	G	E	G	P
Hardness	E	E	E	P
Scratch resistance	G	E	E	G
Chemical resistance	P	E	P	P
Weather resistance	E	P	G	G
Abrasion resistance	G	G	G	P

DEVELOPMENT OF EPOXY POWDERS

Epoxy powders have led the tremendous growth that has occurred in powder usage since 1965. The development of these systems began seriously in about 1960; before that time, there was a very limited use of epoxy powders in the U.S.A. for electrical insulation purposes, but little effort had been devoted to the development of epoxy powders as coating systems on a large scale. However, in the period 1960–64, a considerable amount of research and development was carried out in Europe in order to design suitable powder coating formulations based on epoxy resins and to identify the best methods of manufacturing these powders. Many different groups of curing agents were examined, such as aromatic amines, boron trifluoride complexes, dicyandiamide (DICY) and acid anhydrides and various resins and combinations of them. Among the manufacturing methods investigated were dry blending, ball milling, Z-blade mixing, solvent evaporation and even continuous dispersion.

This work led to the introduction of a number of powders in the paint industry in 1963–64[2-4] and stimulated many paint companies to evaluate powders. This interaction between resin manufacturers, paint manufacturers and the ultimate end-users of the powders led to a closer definition of the type of powder required and identified the limitations of the existing systems, such as long stoving times at relatively high temperatures and yellowing and chalking on outdoor exposure. Faster curing agents were therefore introduced in 1968, which enabled powder systems to be cured in half the previous time at 180 °C and cured at temperatures down to 130 °C. *Figure 6.1* illustrates the effectiveness of these more reactive curing agents and shows the minimum stoving times as

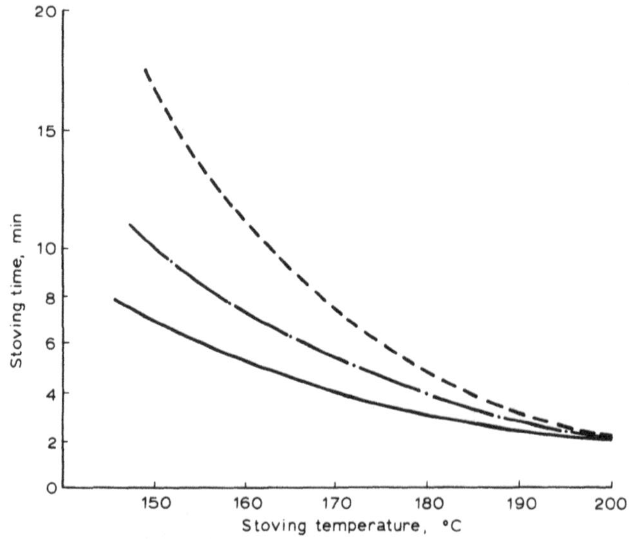

Figure 6.1. Minimum stoving times on aluminium foil[5] *:* – – – –, Epikure 107; –·–·–, Epikure 107–108 mixture; ———, Epikure 108 (Epikure is a Shell registered trade-name) (Courtesy Shell International Chemical Company[5])

a function of temperature for various powders using three proprietary curing agents. The powders were coated on to aluminium foil in order to eliminate the heat capacity of the substrate.

These highly reactive powders offered welcome improvements to the user but caused powder manufacturers to reconsider their methods. A long residence time in a heated mixer would cause the powder to gel before it could be fully blended and discharged into the grinding mill. Consequently, continuous dispersion techniques were re-examined, such as the use of extruders, despite their high initial cost. The current European situation is that both batch and continuous methods of making powders are used in order to prepare a wide variety of systems that possess different reactivities and properties and a colour range comparable with that of normal paint materials.

ECONOMIC ANALYSIS OF EPOXY POWDERS

In the period 1965–70, it has been estimated[6] that the consumption of all coating powders in Western Europe increased from 2000 to 12 000 tons, a six-fold increase in 5 years! Most of this material was applied by the fluidised bed technique, but a rapidly increasing amount was applied by electrostatic spraying. Over the same period and within the totals mentioned, the sales of epoxy powders in Europe grew from zero to 3000–4000 tons[7], and are estimated to become 15 000 tons by 1975.

The situation in the U.S.A. has been different. For some years, epoxy powders were regarded as speciality products and used chiefly for internal and external linings of steel pipes and coatings for certain electrical equipment. However, since 1970, the attitude to powders in the U.S.A. has changed, probably owing to the implementation by different States of clean air and anti-pollution legislation (the well known Los Angeles Rule 66). The U.S. market is therefore now rapidly developing along similar lines to the European market. In 1971, the total U.S. powder market was about 12 000 tons, of which epoxy powders represent about 5000 tons. By 1976, it is estimated that the market will have grown to 60 000 tons, of which 15 000 tons will be epoxy powders. An approximate end-use breakdown of the 5000 tons of epoxy powder consumed in 1971 in the U.S.A.[8] is:

Pipelining	2500 tons
Can coating . . .	500 tons
Electrical	1000 tons
Metal decorative . .	1000 tons

Clearly, the achievement of these forecast large increases in consumption of epoxy powders by 1975 requires market acknowledgement of not only the technical, health, safety and other advantages already discussed, but also clearcut economic advantages. These advantages have been discussed in a number of papers, and derive chiefly from the following technical aspects:

(1) The absence of solvents leads to savings in the cost of solvent removal, elimination of the need to use flame-proofed equipment and lower insurance premiums.
(2) The ability to spray powders to form a thick coat in one application allows the number of spraying operatives to be halved, as only one coat of powder need be applied instead of two coats of solvent-containing enamel.
(3) There will be a reduction in stoving costs (e.g. reduced energy requirement, fewer ovens needed).
(4) Electrostatic spraying can be carried out by inexperienced spray operatives, whereas skilled sprayers are needed for solvent-borne paints.
(5) Any overspray of powders can be recovered and used again, thus ensuring nearly 100 per cent utilisation of the material. Conventional stoving systems frequently have wastage rates of up to 20–25 per cent[9].
(6) Articles coated with an epoxy powder can be machined without damaging the coating. It is therefore cheaper to powder coat components and subsequently machine and assemble them, avoiding expensive masking, rather than coating the assembled components with a conventional paint after the machining operation has been carried out[10].
(7) The excellent physical properties of epoxy powders lead to a reduction in the damage that can occur to coated articles as a result of rough handling during assembly, storage and transport.

Many analyses have been made of the cost of epoxy powder coatings compared with that of conventional solvent-borne systems.

Williamson[11] considered the costs of epoxy stoving enamels and epoxy powders and summarised his findings, which are given in *Table 6.2*, by stating that epoxy powders are competitive with (epoxy) stoving paints where two coats are required to give an acceptable film thickness.

Table 6.2

COMPARATIVE COSTS OF EPOXY STOVING ENAMEL AND EPOXY POWDER (AFTER WILLIAMSON[11])

	Two-pack epoxy stoving enamel		*Epoxy powder*
Application method	*Electrostatic spray*	*Air spray*	*Electrostatic spray*
Solids, % by vol. (RFU)	38	38	100
Practical costs, 0.002-in dry film thickness, p/ft^2	0.88	1.9	0.71
Theoretical costs, 0.002-in dry film thickness, p/ft^2	0.63	0.63	0.54

Ruzicka[12] was also of the opinion that, in general, the cost of an epoxy powder coating was about the same as that of two coats of a stoving enamel system. In terms of material costs needed to cover a mild steel refrigerator shell with a surface area of approximately 25–30 ft^2, Ruzicka calculated the following:

	Material cost, £
(1) Epoxy powder; 0.0025-in film thickness (98 per cent utilisation; coverage 50–60 ft^2/lb at 100 per cent utilisation)	0.375
(2) 1 coat of good quality acrylic; 0.00125-in film thickness (80 per cent utilisation; coverage 500–600 ft^2/gal at 100 per cent utilisation)	0.15
(3) 1 coat of stoving primer; 0.00125-in film thickness (80 per cent utilisation; coverage 500–600 ft^2/gal at 100 per cent utilisation	0.11

The total material cost of the two-coat system (primer + acrylic) is therefore £0.26, a difference of £0.11 in favour of this system compared with the powder coating. But it is necessary also to take into account the savings in application and overhead costs that accrue when using powders, such as those mentioned already in this section (i.e. halving the number of spraying operations, reducing the number of articles rejected, etc.), which will vary between companies, but can at least amount to the difference in raw material costs.

Further cost comparisons have been given by Klaren[13], Moore[14] and Krak and Salzgeber[10], all of whom confirmed the general view that epoxy powder coatings compare very favourably in cost with solvent-borne stoving systems, when all costs involved are taken into account. Naturally, approximations must be made in such calculations, and it is dangerous to attempt to draw too close a comparison between the situations in different countries, but the evidence from all published sources points to the same general conclusion.

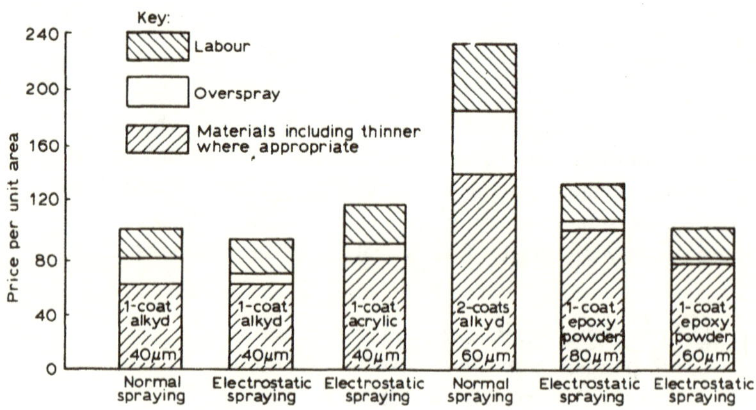

Figure 6.2. Cost comparisons for solvent-borne stoving enamels and epoxy powder coatings (After Klaren[13])

Figure 6.2 is taken from Klaren's paper[13] and shows that a 60-μm layer of epoxy powder is comparable with a 40-μm layer of a stoving alkyd paint applied in one-coat normal spraying. Moreover, a 40-μm coat of acrylic, even when applied with an electrostatic spray, is more expensive than a 60-μm epoxy powder coating. Similarly, *Table 6.3* by Moore[14] considers the relative costs of various systems at different film thicknesses. Powders,

Table 6.3

COST COMPARISON OF SOLVENT-BORNE PAINT AND POWDER COATINGS (AFTER MOORE[14])

Surface coating	*Volume solids,* per cent	*Spraying losses,* per cent	*Desired film thickness,* μm	*Actual film thickness,* μm	*Covering rate,* m²	*Cost per litre,* Swedish Kr*	*Cost per square metre,* Kr*
Alkyl–amino paint	45	40	30	30	9.0	4.35	0.50
			40	40	6.8		0.67
			50	50	5.4		0.85
			60	60	4.5		1.01
			80	80	3.4		1.34
			100	100	2.7		1.69
Acrylic paint	46	40	30	30	9.0	6.40	0.73
			40	40	6.8		0.97
			50	50	5.4		1.22
			60	60	4.5		1.47
			80	80	3.4		1.95
			100	100	2.7		2.45
Epoxy powder	100	3	30	60	16.0	14.85	0.93
			40	60	16.0		0.93
			50	60	16.0		0.93
			60	60	16.0		0.93
			80	80	12.0		1.24
			100	100	9.7		1.53

* 1 Kr ≈ 9.5 p.

although more expensive at lower thicknesses, become competitive at thicknesses of 50–60 μm and above.

Two other sets of costings also bear out this advantage of powders at film thicknesses greater than 50 μm. *Table 6.4* shows that the cost per square metre of the epoxy powder is less than that of the solvent-borne paint for comparable performances, and *Table 6.5* illustrates the lower cost of coating 100 units of an article with an epoxy powder compared with a stove enamelling technique.

Table 6.4

COST COMPARISON FOR EPOXY POWDER AND SOLVENT–BASED PAINTS OF COMPARABLE PERFORMANCE (AFTER *British Plastics*[6])

	Epoxy powder	*Solvent based*
Price, £/kg	1.044	0.58
Solids content, per cent	100	50
Overspray losses, per cent	1–2	20–25
Price per square metre at 50 μm, £	0.078	0.112
Component costs, £/m²		
Energy	0.014	0.007
Chemicals	0.004	0.004
Paint *vs* powder costs	0.047	0.089
Wages	0.005	0.004
Plant	0.008	0.008
Total	**£0.078**	**£0.112**

Table 6.5

COST COMPARISON OF COATING 100 UNITS (AFTER *British Plastics*[6])

Technique		*Cost, £*
Stoved enamel	Primer (2.5 pints at 165 p/gal)	0.52
	Degreasing and working time	3.65
	Top coat (2.1 pints at 164 p/gal)	0.44
	Wages and ancillary costs	1.83
	Total	**£6.44**
Epoxy powder	Powder (1.4 lb at 60 p/lb)	0.84
	(No degreasing)	–
	Wages and ancillary costs	0.86
	Overheads	2.92
	Total	**£4.62**

USES FOR EPOXY POWDERS

Clearly, epoxy powders are well prepared to obtain a large part of the conventional stoving enamel market, not only on economic grounds but increasingly on the need to avoid atmospheric pollution by the venting of solvents into the atmosphere. There are, of course, many other advantages in using powders, and some of these are illustrated by the case studies mentioned below.

A typical example of the way in which powders are replacing solvent-containing paints is in the finishing of parts for a range of steam-powered models. A high-quality finish to the models is essential and the several small components are difficult to spray. Hand spraying was slow, the output being a few hundred per hour and, in addition, 50 per cent of the paint was wasted in overspray, two coats were needed on each side of the wheels and were applied and stoved separately, and the reject rate was high. The decision was taken to use an epoxy powder coating applied by electrostatic spray. The advantages resulting from this decision included a six-fold increase in output, a reduction in overspray wastage to 5 per cent, a reduced reject rate and the elimination of the need for a priming coat, one application of the epoxy coating being sufficient. Moreover, the powder produced a better and more durable finish.

Most of the powder used at present is for the decorative coating of small articles such as kitchen and bathroom equipment, motor vehicle parts, fire extinguishers, springs, bicycles and tubular steel furniture. However, this list is now being extended to include larger items such as central heating units and refrigerator shells and powders are now being used in an ever-widening range of decorative and industrial finishes. Three further case histories will serve to illustrate these points.

Ascot gas water heaters were previously coated with vitreous enamel, and this coating influenced the overall shape of the appliance[15]. When wishing to change the design and shape of the heater to one which possessed a rectangular outline and square corners, it was foreseen that there would be extreme difficulties in obtaining a satisfactory vitreous enamel finish on the corners and edges. It was therefore decided to replace the vitreous enamel by an epoxy powder coating which would have the following advantages:

(1) Even application by spraying and satisfactory cover for edges and small radii.
(2) A smooth hard attractive finish would be obtained, resistant to chipping and damage.
(3) Only one coat of powder would be required.
(4) There would be no smell or fire hazard involved in using the powder.

A manufacturer of industrial electrical outlets sockets and plugs changed the coating system from a solvent-borne paint to an epoxy powder[16] because of:

(1) The difficulty in obtaining trained labour to spray solvent paints satisfactorily.
(2) Easier plant maintenance with powders.

It was also found that the overall cost of using powders was no greater than that of using solvent paints.

A manufacturer of cold storage plant normally used a solvent-containing high-bake acrylic coating to give his equipment a hard, white finish. However he changed his entire production over to a white epoxy powder coating[17] and achieved the following advantages:

(1) Previously, of the two different types of steel used in the plant, one required an etch primer. With the epoxy coating, no primer was needed for any of the steel.
(2) Material wastage was reduced to 2 per cent.
(3) There was virtual elimination of rejects.
(4) Throughput was greatly increased.
(5) The labour content of the work was cut by 50 per cent and shift work ended.
(6) Product quality and consistency improved.

Only a small proportion of the epoxy powder used is for items that require a high degree of protection against corrosion or chemical attack[12], whereas when the powders were first introduced it was expected that their main uses would be as chemically resistant coatings for tanks, pipelines and other similar items. However, coating of pipelines with epoxy powders is an important use and is discussed in Chapter 5. Because of their electrical insulation properties as well as their other advantages, the powders are used to coat busbars, transformer housings, etc., and to provide insulation in small h.p. motors.

THE FUTURE FOR EPOXY POWDER COATINGS

One of the largest future potential markets for powders is in automotive coating, with the aim of applying the top finish to the car as a single coating of the powder. Already in the U.S.A., Ford and General Motors have plants in prototype operation for powder coating top body finishes[8] and in Europe Citroen in France and Fiat in Italy have plants applying an epoxy powder primer to automobile chassis[6]. In the U.K. Rover has been coating rocker boxes for some years and Leyland National buses also use a similarly coated part[6]. Epoxy primers and primer surfacers have long held a dominant position in automotive finishing and should retain this position if powders are adopted for primers, but it is unlikely that epoxy systems will be used for top coats because of their chalking and lack of gloss retention. No doubt acrylic or polyester powders are likely to be accepted for this purpose and the U.S. plants mentioned above are reported to be using a thermoplastic acrylic powder[8].

Subject to the limitations of yellowing after continuous exposure to elevated temperatures and chalking on exterior exposure, epoxy powders will continue to penetrate the decorative and industrial stoving alkyd market. The wider adoption by Governments of anti-pollution legislation will also help to maintain the switch to solventless coatings. There is no doubt that the traditional outlets for epoxy systems such as internal and external anti-corrosive coatings for pipes and particularly automobile primers will become important markets for the powders. More speculatively, metal coil coatings for wall claddings, caravans, etc., are likely to be based on epoxy powders and experimental marine powder coatings have already been applied on trial sites in Japanese shipyards. The future market for epoxy powders therefore seems bright but there will be increasing pressure in this market from the manufacturers of acrylic and polyester powders.

REFERENCES

1. STEPHENS, T. J., *Surface Coatings,* **1**, No. 3, 90 (1965)
2. *Ind. Finish.,* **16**, No. 193, 29 (1964)
3. *Paint Mfr,* **34**, No. 8, 38 (1964)
4. WEBER, J. R. and SALZGEBER, R., *Paper to 7th FATIPEC, Vichy,* 1964

5. *Rapid Curing Epikote Resin Powder Coatings,* Shell International Chemical Company
6. *Br. Plast.,* **44**, No. 12, 45 (1971)
7. VERHULST, G. P. J., *Ind. Finish.* **23**, No. 278, 14 (1971)
8. *Mod. Plast. Int.,* **2**, No. 5, 48 (1972)
9. HUMPHREY, K. W., *J. Oil Colour Chem. Ass.,* **476**, 434 (1964)
10. KRAK, H. and SALZGEBER, R., *Surface Coatings,* **6**, No. 2, 43 (1970)
11. WILLIAMSON, R., *Thermoset Powder Coatings,* WMS (ENG), Feltham, Middlesex, 66 (1968)
12. RUZICKA, F. C. J., *Thermoset Powder Coatings,* WMS (ENG), Feltham, Middlesex, 29 (1968)
13. KLAREN, C. H. J., *Amer. Paint Jour.,* **55**, 64 (1971)
14. MOORE, W. V., *Lecture to 1971 Oil & Colour Chemists Assoc. Conf.*
15. *Prod. Finish.,* **25**, No. 5, 15 (1972)
16. *Prod. Finish.,* **25**, No. 2, 35 (1972)
17. *Ind. Finish.,* **24**, No. 290, 28 (1972)

SECTION III
Electrical and Electronic Insulation

Introduction

The factors to be considered when selecting materials for electrical insulation purposes have been described most succinctly by the late Willis Jackson[1]:

> The primary function of the electrical insulation of an equipment is to confine the current to the conductors constituting its electrical circuit. In order that it may do this at least some part of it must serve to provide mechanical support for these conductors within the structure of the equipment. For this reason, the mechanical and physical properties of insulating materials are usually of dominant importance in determining their individual suitability for particular applications. Moreover, good electrical insulating materials are almost invariably good thermal insulators. The provision of an electrical barrier thus inevitably introduces a thermal one. The significant electrical and mechanical properties are therefore not the ambient temperature ones but those at the elevated temperature of operation, maintained over long periods, which this thermal condition imposes. Since generally speaking these properties deteriorate with rise of temperature, though to varying degrees with different materials, it is necessary to limit the maximum operating temperature rise above the ambient . . . . Further complication bearing on the above arises from the atmospheric conditions in which the equipment is operated, in particular from the effects of moisture absorption, oxidation processes and attack by chemical reagents. In other words, it by no means suffices to choose materials and design insulating systems which will meet the specified electrical requirements initially; they must continue to meet these requirements over a very long period of service.

Very many kinds of materials are used as insulators, ranging from oils, gases and synthetic and natural polymers to glasses, minerals, salts and ceramics. All are characterised by having a scarcity of free electrons. While they can and do conduct an electric current when subjected to a voltage, the currents are extremely small and even then a very high voltage is required in order to produce them. Insulators can therefore be regarded as conductors with very high resistivity, so high, in fact, that the current they pass can be neglected in power equipment but not in electronic equipment. There is no clear dividing line between conductors and insulators, but in general it can be said that conductors have a volume resistivity of between 10^{-6} and 1 ohm cm, poor conductors between 1 and 10^6 ohm cm and insulators between 10^6 and 10^{20} ohm cm.

In the early days of the electrical industry, insulation was achieved by using materials that consisted almost wholly of processed forms of natural products. These were typically mica, glass and porcelain or, when flexible materials were needed for wiring, cotton and silk[2]. Rubber was also used in its flexible form as a wire covering and, after vulcanisation to ebonite, for mouldings. Wire enamels initially based largely on phenolic resins were developed as a means of reducing the volume occupied by rubber or textile coverings. Moreover, enamelled wires enabled more turns to be wound per unit volume on, for example, the stator coil of a dynamo. This enabled higher flux densities to be obtained together with higher working temperatures and hence increased machine efficiency.

Mica tape and sheet (mica folium) used shellac or modified bitumens in order to bond the mica to the support, and transformers and other equipment are often still insulated and protected from moisture by coating or embedding in bitumen. For small components, such as the condensers used in radios, paraffin waxes are used as insulants. However, both bitumen and wax progressively soften and flow as the component becomes warm and, in general, are mechanically weak.

As electrical equipment was required to operate at higher temperatures and increasing efficiency, often in aggressive environments, the need for improved insulation grew. The industry therefore turned towards synthetic polymers as improved substitutes for the natural products that had been used satisfactorily for many years. The earlier synthetic resins, such as the phenolics and later the polyester resins, were both widely used for electrical insulation in the late 1940s, although phenolics had first been used much earlier. Unfortunately, both groups of resins suffer from certain limitations in their use as far as electrical insulation is concerned. Phenolic resins are, of course, suitable for the production of mouldings and laminates, but are not chosen for casting and encapsulation purposes because of their high shrinkage on cure (8–10 per cent by volume), and evolution of volatile materials during cure and their high power factor, which gives rise to large power losses and consequently to overheating. Polyesters also suffer from high shrinkage and have poor dimensional stability, although they offer reasonable electrical properties. Silicone resins are fairly expensive but offer good insulation and dielectric properties that are maintained at elevated temperatures. Unfortunately, their mechanical properties are poor, as is their adhesion to different substrates.

Epoxy resins, although they do not possess such good electrical properties as the silicones, are nevertheless superior in this respect to phenolic and polyester resins. Amongst the properties which recommend epoxy resins for electrical power applications are:

(1) The ability to cure with negligible shrinkage and absence of volatiles.
(2) High mechanical strength, which is retained at elevated temperatures.
(3) Good dielectric strength, volume and surface resistivity.
(4) High resistance to chemical and atmospheric attack and to absorption of water.
(5) Good heat resistance and dimensional stability.
(6) Remarkable adhesion to metals, glass, porcelain, rubber and other materials.
(7) The ability to be cured at low temperatures when required, e.g. encapsulating heat-sensitive components.

These properties are allied with the ability to use epoxy resins in a variety of manufacturing methods and techniques. By using a casting process, complicated shapes can be formed in one operation and insulation cast around coils and conductors. For large equipment, such as transformers and switchgear, cast insulation enables the use of a number of separate components to be avoided and processes such as the laborious hand application of tape to be eliminated. Simpler and more compact designs can often be adopted because irregularly shaped conductors can be easily embedded in the resin, which can save space and reduce costs while providing good insulation and protection against the environment. The resin formulations can be spread into a thin film in a bonding process or used in a dip-tank for impregnating components. Epoxy resin-based coatings are applied by dipping, spraying or in powder form and small components can be encapsulated by means of a transfer moulding process. Finally, the resin can be used as the matrix in the preparation of laminates, typically those in which glass-fibre is used as reinforcement.

Filament winding, low- and high-pressure moulding and the use or pre-impregnated cloth are techniques in which epoxy resins are used. Epoxy resins, with their ability to offer a nearly unique set of properties and to be used in this wide range of fabrication techniques, have brought new concepts of design to the electrical industry and given design engineers a new degree of freedom in the production of intricate high-performance components. Moreover, with high-voltage equipment in particular, the contrast between modern designs and those used before the advent of epoxy resins is sufficiently marked to justify regarding the date of the introduction of cast epoxy insulation as a turning point in electrical technology. Epoxy resins have also served to encourage advanced manufacturing techniques with accompanying increases in productivity and decreases in production costs.

REFERENCES

1. JACKSON, W., *The Insulation of Electrical Equipment,* Chapman and Hall, London (1954) Chap. 1
2. WARREN, H., *Electrician,* **110**, 484 (1933)

7
Rotating Machines

Motors and generators with a wide range of outputs are produced, the smallest output being a fraction of a horse power and the largest several thousand horse power. Operating voltages also vary from a few volts to over 20 000 volts. These many different types of rotating machines, from the very large turbo-alternators to the very small motors in, for example, mini-tape recorders, lead to a varied requirement for insulating materials.

The function and properties of insulation in rotating machines

The primary function of insulation is to separate circuits and conducting parts that have different potentials. The common technique is to apply a dielectric barrier of a solid structure or a continuous insulating film possessing an intrinsic dielectric strength well above that of an equivalent air gap.

Insulation used in rotating machines can be classified as follows:

(1) *Ground insulation* is the major insulation between the winding and the core, and it must be capable of withstanding the full voltage applied to the machine. The most common and economically attractive technique used in ground insulation is to place a sheet of insulation within the slot of a steel core (slot-liner). An alternative method is to wrap a sheet of dielectric material to the straight portion of a coil, to be inserted in the slot, with an overlap beyond the core forming a tube with a high intrinsic breakdown strength. Formed coils designed for high-voltage operation are wrapped throughout including over-hangs. The third, most versatile, technique used to insulate the coils from the core is to use taped insulation.

(2) *Turn insulation* is used in order to prevent short-circuiting between turns passing through the same slot. The insulation can be in the form of enamel, fabric or splittings such as mica; end-turn insulation is taped.

(3) *Phase separation insulation* is used on a.c. machines because the frame is usually at an intermediate voltage with respect to the terminals of the windings. Therefore, the voltages between the phases will be higher than to the ground and insulation is provided between the phases both in the slots and between the end windings. Laminates are generally used in the slots and insulating sheets and tapes for the end windings.

(4) *Commutator insulation* is provided between commutator segments of d.c. machines and a.c. commutator machines and ground insulation between the commutator segments and the shaft.

(5) *Core insulation* is used between thin sheets of steel used to build stator and rotor cores.

(6) *Cords and lacing tapes* are used for structural and finishing purposes. Their use is mainly mechanical but they serve as a base for waterproofing provided by impregnation.

Requirements of insulating materials

An ideal insulation material would have a high dielectric strength, high electrical resistivity, low dielectric loss, good resistance to moisture, good heat conductivity, good mechanical properties and good thermal stability. Considering the last property, the temperature of various parts of rotating electrical machines increases due to heat generated by various losses, as follows:

(a) Rotor winding losses.
(b) Stator winding losses.
(c) Iron losses.
(d) Friction and windage losses.
(e) Dielectric losses.

An increase in temperature above the design limits in rotating machines leads to:

(1) Differential expansion between the insulation, the surrounding steel and the conductors, which gives rise to mechanical stresses and subsequent abrasion of the insulation.
(2) A reduction in the electrical properties.
(3) A reduction in the mechanical properties.
(4) Chemical changes in the insulation, which usually change the electrical properties.

Obviously, the operating temperature is of fundamental importance to the designer in the selection of insulating materials. Because of this importance, an internationally recognised temperature classification is in use (*Table 7.1*).

Table 7.1

TEMPERATURE CLASSIFICATION OF INSULATING MATERIALS

Class of insulating material	*Maximum temperature, °C*	*Equivalent ASTM standard*	
		Class	*Maximum temperature, °C*
Y	90	O	90
A	105	A	105
E	120		
B	130	B	130
F	155	F	155
H	180	H	180
C	above 180	C	220
		over C	> 220

When designing machines, it is therefore essential that a material should be intended to operate within its temperature limit and, in addition, that it is not affected during the operation of the machine by neighbouring materials that are at different temperatures.

EPOXY INSULATION IN ROTATING MACHINES

The techniques and problems involved in providing machine insulation vary according to the size of the machine, and it is therefore convenient to consider separately two types of machine, large motors and turbo-alternators of perhaps 50 h.p. and upwards, and the small machines with lower ratings. However, in both groups of machines, epoxy resins are used chiefly to provide classes B and F insulation. Class E insulation is of minor importance for epoxy resins, as improved polyesters have largely replaced them in the insulation of small motors, which is the chief area of class E applications. However, epoxy powder coatings and wire enamels are used in small motors, and the future for powders is bright. Many small motors are also totally encapsulated in epoxy resin for protection against adverse environments.

In class B insulation, epoxy systems are used in composites as combinations with mica, glass and terylene for ground insulation, wrappers, spacers and end-turn insulation. They are particularly valuable for high-tension applications. In class F insulation, epoxy resins are used chiefly in ground insulation in large motors.

The following sections deal separately with large and small rotating machines and the reader will note that the treatment of large machines is mainly technical, whereas that of the small machines is broader in scope and largely applicational. The reason for this difference in approach lies in the nature of the two subjects. Large machines are concerned almost exclusively with power generation and epoxy resins are used chiefly in epoxy–mica tapes and epoxy–glass laminates for slot-liners. In contrast, small rotating machines are used in many different types of equipment and epoxy insulation can be introduced by various techniques, such as vacuum casting, dipping and powder coating.

Large rotating machines

Epoxy insulation has become well established in machines of over 50 h.p., traction motors and turbo-generators, and by the mid 1970s it is almost certain that epoxy–mica systems will become their standard form of insulation, as has probably already occurred in Germany and Switzerland. The reasons are as follows.

The continued growth in physical size and output of rotating electrical machinery, particularly turbo-generators[1], has accentuated the problem of heat dissipation and the ability of insulation systems to withstand the increasingly severe thermal, mechanical and electrical stresses placed upon them. Designers have been required to keep the physical size of the machine to a minimum and to pay ever-increasing attention to the reliability of the machine and therefore to the reliability of its insulation. In this regard, it is of prime importance that the insulation system should possess adequate mechanical properties as well as acceptable electrical properties. Experience has shown that the few failures of large machines that have occurred have been due more to mechanical faults than to outright electrical failures of the insulation[2,3].

The coil insulation systems used in large rotating machines in all countries are based almost exclusively on mica in the form of paper reinforced with synthetic or glass-fibre tapes. The earliest machines used mica or varnish-impregnated cloth insulation, the mica in the form of flakes (mica splittings) being supported on a tape or sheet of paper or textile material and being bonded together by shellac or bitumen. As machine sizes increased, shellac came to be regarded as too brittle and was replaced by bitumen, which was often slightly hardened by the addition of drying oil. These mica tapes and sheets have been very successful in use and have been employed widely throughout the whole range of both a.c. and d.c. rotating machines, including turbo-generators up to 500 MW capacity operating at voltages up to 22 kV. However, difficulties have occurred under operating conditions owing chiefly to the thermoplastic nature of the bitumen binder. Differential thermal expansion and contraction of the stator copper, iron and insulation, brought about by the difference in temperature in the machine when idling and at full load, can cause stretching of the coil insulation tape with a consequent reduction in insulation thickness. Breakdown of the insulation is then likely to occur. In addition, thermoplastic insulation materials are inclined to swell with time, particularly as the thermal and mechanical forces to which they are being subjected are increased.

Machine designers are now attempting to provide more output for a given frame size, which requires more copper in the slot and higher operating temperatures for the machine. The insulation system must therefore withstand these higher temperatures and provide electrical insulation with a smaller wall thickness. The traditional method is also being superseded because of the rapidly increasing cost of split mica and of making it into sheets by hand operation. Moreover, an economic improvement can be shown in the ability to remove a thermosetting material from a press maintained at constant temperature compared with the need to cool presses, which is essential with the traditional thermoplastic materials.

Attempts were therefore made to replace bitumen by a thermosetting resin as the binder for mica. Phenolic resins were considered but found to be rather brittle. Attention was then turned to polyesters, which were found to give considerable improvements over bitumen. In particular, they enabled an insulation system to be fabricated that did not swell radially when in use or lose its mechanical strength when hot. The low viscosity of uncured polyesters was also advantageous, as it allowed easy impregnation of mica. However, one limitation to their use was their short pot-life, which restricted the storage life of tape or sheet impregnated with the resin.

In the 1950s, epoxy resins were investigated as binders for mica and found to offer improvements in strength compared with polyesters but possessed an unacceptably high viscosity, which prevented them from being used for impregnating processes. Considerable development work was then carried out, chiefly by electrical companies in the U.K. and other European countries, aimed at producing suitable epoxy formulations for use with mica. At about the same time, mica paper was introduced, which consists of small mica platelets made into a type of paper by using modified paper-making machinery. Development work has continued since the early days of epoxy resin-mica systems and there are now available to the insulation engineer many different epoxy resin formulations that are used with reinforced mica paper in tape or sheet form. This insulation is applied to machine coils by two main methods, which have been described in detail in numerous papers[3–8], and are basically as follows:

(1) The impregnation method, in which the mica paper and its backing is bonded with a minimum amount of epoxy resin and this dry paper is wrapped around the machine coils. After rigorous vacuum drying, the wrapped coil is subjected to vacuum–pressure impregnation by immersion in a solventless epoxy resin system. After the coil has been impregnated, the resin is cured and the insulation pressed into shape in a heated mechanical press.
(2) The pre-impregnation (resin-rich) method, in which the mica tape or sheet plus its backing is pre-impregnated with an

epoxy system in a similar way to the preparation of 'pre-pregs' based on glass-cloth. If the epoxy system contains solvent, as is usual in the manufacture of glass 'pre-pregs', as much as possible of it must be removed by the normal hot vacuum process prior to applying the tape to the coils. Alternatively, a solventless epoxy system can be used, which removes the need for the solvent removal step[9]. When the coils have been wrapped with the 'pre-preg', they are consolidated into shape and the resin is cured by the application of heat and pressure.

It is not intended to compare the merits of these two processes; both are widely used and are well established techniques. However, the impregnation method requires a higher capital investment in plant compared with the pre-impregnation method, provided that the user of 'pre-pregs' either employs a solventless system or purchases his mica 'pre-pregs' elsewhere.

Clearly, epoxy resin binders for mica insulation systems are now well established and have no doubt reached this position because of their superior all-round properties compared with other commercially available resins. In particular, epoxy resins offer excellent adhesion to copper and other materials, superior mechanical properties to most other thermosets and good electrical properties. In addition, the wide choice of resins and curing agents enables an epoxy formulation to be tailor-made to meet most individual requirements. Some comparative properties of epoxy- and bitumen-bonded mica insulation systems are given in *Table 7.2.*

The use of epoxy-bonded mica paper has allowed machines to be designed up to full class F temperatures (155 °C), whereas class B (130 °C) was the previous limit. The improved quality in materials and application methods has also enabled designers to reduce the thickness of the machine stator wall insulation by up to 20 per cent, thus allowing a smaller, more efficient machine to be produced for the same kVA rating[10]. The use of epoxy–mica 'pre-preg' tape is being extended to coils up to 30 kV, which may also lead to further reductions in wall thickness and hence increase still more the amount of copper in the slot.

In the future, it is probable that the present systems will remain in use, any improvements being processing speeds increased while the present properties and stability are retained. Any new system

Table 7.2

COMPARISON OF EPOXY- AND BITUMEN-BONDED MICA INSULATION SYSTEMS

Property	*Epoxy–mica*	*Bitumen–mica*
Longitudinal coefficient of expansion	9×10^{-6}	10×10^{-6}
Thermal conductivity, cal/cm/s/deg C	7×10^{-4}	6.9×10^{-4}
Flexural strength, p.s.i.	21 000	2200
Flexural strength, MN/m^2	147	15
Flexural modulus, p.s.i.	2.2×10^6	0.25×10^6
Flexural modulus, GN/m^2	15	2
Compressive strength at room temperature, p.s.i.	41 500	40 000
Compressive strength at room temperature, MN/m^2	290	280
Compressive strength at 80 °C, p.s.i.	37 500	1870
Compressive strength at 80 °C, MN/m^2	262	14
Dielectric constant	4.8	3.5
Electric strength, 50 Hz 60 s (300 mil thickness), V/m	425	350

would need to show very worthwhile technical improvements at existing cost levels before manufacturers would consider changing the now fully established resin systems.

In addition to epoxy–mica insulation, laminates based on epoxy resin and glass-cloth are also used extensively in electrical machine insulation for components such as slot liners and slot wedges of various shapes and sizes. This form of insulation requires a material that has good mechanical strength and dimensional stability, good insulation properties and the capability of operating up to class F temperatures (155 °C). Stator restraining rings up to several feet in diameter are also fabricated from epoxy laminates, usually made from rovings by filament winding or from wet-wound tape.

Small rotating machines

The designers of smaller generators and motors have also been confronted with the need to decrease costs, increase power to weight ratios and improve the reliability of equipment in order to reduce maintenance costs. Higher power to weight ratios can be achieved by a reduction in machine size while maintaining the power output, but this usually requires more efficient dissipation of the heat generated during motor operation. Alternatively, increased power can be obtained, at constant weight, by increasing the permissible temperature of the windings, that is, by using higher current densities in the winding copper. The extent to which the current density can be increased is dependent upon the ability of the insulation to withstand the temperature generated, provided that the other elements, such as bearings, etc., are not adversely affected.

A reduction of 5–10 °C in temperature rise can be expected by impregnating motors with solventless varnishes[11], which eliminate all air and solvent voids inside the windings or between slot insulation and core, which may act as thermal insulation barriers to the passage of heat. The availability of epoxy resin insulation varnishes which retain their integrity at elevated temperatures and which can be applied by a variety of techniques has been a major factor in enabling rotating machines to be made that operate at class F temperatures (155 °C). This has already been mentioned in the previous section.

A further very important use for epoxy resins is in the protection of stator windings against the ingress of moisture and contaminants and against adverse environments by the complete encapsulation of the windings. Such treatment has led to great improvements in the reliability of equipment, with consequent reductions in maintenance costs.

Most major electric motor manufacturers throughout the world now offer a range of epoxy-encapsulated motors which are typically used in chemical plant where wet conditions predominate, in domestic and industrial appliances such as washing machines and dish-washers, and dusty or contaminated atmospheres, for example those obtained in carbon black manufacture or where fine abrasive powders are used[12]. As an example of this last situation, a totally enclosed fan-cooled motor was not giving good service when installed in a fertiliser plant. It was found that the motor was drawing in fine particles of fertiliser, which, in humid air, formed acid and destroyed the motor insulation. Replacement of the motor by one with its windings encapsulated in epoxy resins completely solved the problem.

Replacement of a totally enclosed motor that was required to drive an acid pump in a chemical plant by one encapsulated in epoxy resin led to the elimination of motor failure that had occurred in the past owing to acid fumes and moisture degrading the motor insulation. A corn processing plant in the mid-west of the U.S.A. had a 10 h.p. fan motor mounted in a steam duct which broke down at 3-weekly intervals. When impregnated and capped with an epoxy resin, the motor functioned satisfactorily for several years without problems. In many countries, the health authorities require machinery in food processing plants to be scrubbed regularly with caustic solutions, and this treatment necessitates the annual rewiring of many motors. By using epoxy-coated motors with encapsulated field coils, the life of such machinery is greatly extended.

On a.c. equipment, a phenolic varnish provides protection for about 1 min if flooding with seawater occurs while the motor is operating. In contrast, standard production line epoxy-encapsulated motors have served satisfactorily for periods greater than 500 h when immersed in salt water[13].

Traction motors form a particularly important group of electrical machinery for which problems of increased power to weight ratio and increased motor reliability are of prime concern. Moreover, of all traction motors, the self-ventilated, axle-hung type is subjected to the most extreme conditions[14], as it must withstand vibration, shock loads, ice, snow, water, dust, oil, grease and extremes of temperature and it is subject to intermittent loading, which can produce temperature high-spots for short periods, and it must also operate reliably for long periods without maintenance. Epoxy resins are widely used in the insulation systems of these motors for coil insulation and bonding them to their poles because of their excellent bonding properties, mechanical strength and toughness, plus their ability to be formulated as solventless varnishes that can be readily combined with mica or asbestos paper. The results of a series

Table 7.3

TESTS ON RESIN–MICA SUPPORTED INSULATION SYSTEMS FOR FIELD COILS OF TRACTION MOTORS (AFTER KUCHARSEY AND KELLEY[15])

Resin group	*Total coils tested*	*Failed coils*	*Failures,* per cent
Polyester	1300	585	45
Silicone resin	248	94	38
Silicone compound	230	230	100
Epoxy 'A'	575	190	33
Epoxy 'B'	210	23	11
Epoxy 'C'	720	9	1.3

of tests on various resin–mica supported insulation systems for field coils of traction motors have been given by Kucharsey and Kelley[15], and are shown in *Table 7.3.*

Kucharsey and Kelley also reported that the inspection of epoxy-impregnated field coils that had been in service for 6 years did not reveal any breakdown of the resin or the insulation. The effect of various synthetic resins on traction motor cooling was also determined, and the results shown in *Figure 7.1* indicate the effectiveness of

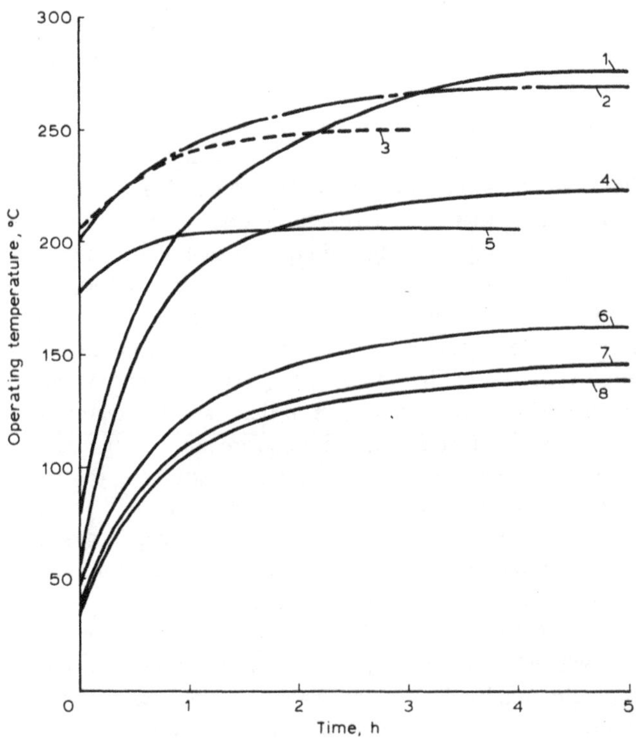

Figure 7.1. Effects of various synthetic resins on motor cooling at 'full load' operation of traction motors: Field coils moulded to pole pieces and treated with synthetic resins. 1 = Conventional insulation; 2 = polyester; 3 = polyester; 4 = silicone compound; 5 = epoxy resin; 6 = epoxy resin; 7 = epoxy resin; 8 = open-face coil (no resin). (From Kucharsey and Kelley[15])

epoxy-impregnating systems in displacing air and filling all the voids between the windings and in the glass-cloth supporting the mica.

The rehabilitation of traction motor field coils in which thermal ageing and other stresses have caused embrittlement and degeneration of traditional insulation materials based on mica, organic binding material, shellac, etc., can also be carried out by using epoxy resins, with significant cost savings. Taking care not to disturb the old insulation, which crumbles readily, a solventless epoxy varnish can be used to impregnate and consolidate the entire system to a void-free mass. The alternative would be to rewind completely the entire coil. About 500 d.c. traction armatures were treated with epoxy resin in this way over a 2-year period, and a failure rate of 1.01 per cent was obtained. Approximately the same number of coils, over the same period, were treated with a phenolic varnish, and showed a failure rate of 15.5 per cent[15].

Motor windings are impregnated for protection against the ingress of dirt and moisture, to improve heat dissipation and to consolidate and hold the windings in place. A number of techniques have been used in order to achieve satisfactory impregnation. Formerly, the windings were dipped into a solvent containing varnish and then heated in a conveyor-type oven in order to remove the solvent and cure the resin. Phenolic resin varnishes were used and the early epoxy systems also contained solvents to reduce the varnish viscosity. During recent years, solventless epoxy and polyester varnishes have been developed that have the advantage of requiring no solvent removal, and therefore minimising waste and removing the possibility of air or solvent voids being formed inside the windings or between slot insulation and core.

However, this market for epoxy varnishes is falling, because for most uses polyester-impregnating varnishes provide completely adequate insulation at a lower price than epoxy insulation. Moreover, it is likely that this trend towards the use of polyester insulation will continue. Both epoxy systems and polyesters are applied by dipping, by the normal vacuum–pressure impregnation bath process or by trickle impregnation, and large numbers of small motors are readily processed.

For the production of sealed motors, protection with epoxy systems is preferred, and the windings are encapsulated by using a simple mould and applying the resin compound by simple hand or vacuum–pressure casting.

A departure from liquid epoxy systems, which will possibly lead to major economies in the production and maintenance costs associated with fractional horse power (f.h.p.) motors, is the use of an epoxy powder coating as the integral insulation layer in stators and rotors. The pre-cut paper or plastic slot liners normally used are replaced with the epoxy coating, on to which the windings are laid directly. This technique removes the need for insertion equipment or hand operation for the paper slot liners. In addition, the amount of wire used in the motor can be substantially reduced, by up to 30 per cent, as the windings do not need to stand off the end-faces of the stator as in the orthodox insulation method and can be wound tightly against the face of the lamination. This reduction in the amount of wire in the motor also reduces power losses, thus enabling the motor to be uprated. Other advantages follow, such as a reduction in the size of the motor housing, a reduction in lamination noise level, the possible dispensing with rivetting or welding of stator lamination stacks, the resin coating serving to band the layers together, and better heat conduction from the windings.

In order to take full advantage of the powder coating technique, it is necessary to effect radical changes in the design of the motors themselves. For example, the shapes of the individual winding slots could be varied, thus varying the number of turns contained in them, and these could be arranged so as to maximise the electrical and magnetic circuit characteristics. This type of configuration can also reduce the amount of unused space in the motor and bring about savings in the cost of laminations.

It is claimed that with a short lamination stack, with its 30 per cent reduction in the amount of wire used, an overall saving of 25 per cent in the weight of a 1/4 h.p. motor can be made. Other advantages in using powder coatings are a reduction in the stock of slot liners that a manufacturer needs to hold, a reduction in motor reject rates owing to the more reliable insulation, and increased flexibility in production given by the ease of change-over between different types of motors.

A further use for epoxy powders is for impregnating the windings of motor coils[16]. Provided that the wire gauge is not too fine, complete saturation is possible. The wire is heated by resistance heating and dipped into a fluidised bed of the powder, which is converted into a low-viscosity liquid when it comes into contact with the hot wire, and this liquid flows between all the windings by capillary action.

The powder coating of armature and stator stacks is carried out on a mass production basis using the fluidised bed technique[17].

Equipment is available that can coat 2000 armatures per day, having a cycle time of about 15 s per component, depending on the size of the armature[18]. Other facilities exist with ultimate capacities of 10 000 armatures per 24-h day[19], which can be used for portable electric products such as dictating machines, record players, tape recorders and movie cameras.

Looking to the future, by the mid-1970s it is probable that because of the trend to higher temperature working, over 50 per cent of all 'small' rotating machinery will be in class B and a further 25 per cent in class F. These are the ratios that apply to classes E and B, respectively, at present. The trend towards the use of polyester varnishes will also continue, as they offer acceptable performance at prices below that for epoxy systems. For sealed motors, however, epoxy encapsulation will remain the preferred method. In fractional h.p. motors, the advantages of using epoxy powders could lead to the widespread adoption of this method of insulation.

REFERENCES

1. *Engineering, Lond.*, **203**, No. 5270, 628 (1967)
2. DOUGLAS, J. L., LACOTTA, M. J. and SMITH, J. W. R., *BEAMA elect. Insulation Conf. Proc.*, (1970) Paper 1c.
3. SEXTON, R. M., *Insulation, Lake Forest*, **74**, No. 4, 56 (1968)
4. FARMER, E., *BEAMA elect. Insulation Conf. Proc.*, (1970) Paper 1.
5. HOYLE, E. L., *BEAMA elect. Insulation Conf. Proc.*, (1970) Paper 1a.
6. MOISEMANN, H., *Insulation, Lake Forest*, **9**, No. 10, 45 (1963)
7. NEAL, J. E., *9th IEEE Insulation Conf.* (1969) Paper 69C33EI-4
8. NYLUND, K. and MOELLER, W., *Bull. Oerlikon*, No. 377/378, p.11
9. ERDMAN, D. E. and LAUROESCH, H. C., *9th IEEE Insulation Conf.* (1969)
10. PRICHARD, G., *Elect. Rev., Lond.*, **187**, No. 24, 839 (1970)
11. POSWICK, J. R., *BEAMA elect. Insulation Conf. Proc.* (1970) Paper 2.
12. LAWTON, C. D. and OSTERLAND, F. C., *Allis Chalmers elect. Rev.*, **23**, No. 2, 32 (1959)
13. LEE, H. and NEVILLE, K., *Handbook of Epoxy Resins*, McGraw-Hill, New York, 17–37 (1967)
14. AUSTIN, G. N., *BEAMA elect. Insulation Conf. Proc.* (1970) Paper 3b.
15. KUCHARSEY, J. J. and KELLEY, W. E., *3rd IEEE Insulation Conf.* (1962) Paper 62-270
16. SMITH, M. J. J., *Thermoset Powder Coatings*, WMS(ENG.), Feltham, Middlesex, 42 (1968)
17. SLATER, D. L., *Plast. Technol.*, 8, No. 3, 28 (1962)
18. *Ind. Finish.*, **16**, No. 198, 36 (1964)
19. *Insulation, Lake Forest*, 8, No. 9, 24 (1962)

8
Distribution Switchgear and Transformers; Cable Jointing

The applications considered in this section are mainly characterised by the term 'heavy electrical casting' and are in the range above 5 kV, which chiefly involves switchgear for electricity distribution and measuring transformers. Epoxy resins are used either as castings or as laminates with glass-fibre in the manufacture of insulators, bushings, transformer insulation and various other switchgear components. These were some of the earliest uses for epoxy resins, being first developed in Switzerland in the late 1940s.

SWITCHGEAR

Until the mid-1950s, the main insulation materials used in switchgear were oil, porcelain, air, resin-bonded paper and phenolic mouldings. None of these materials, either alone or in combination, had the desirable properties of an insulant, and their use in a combined system imposed great restraints on design and caused equipment to be much more bulky than has since been shown to be necessary.

The introduction of cast epoxy resin insulation changed the situation markedly. Epoxy resins have the important characteristics of good electrical insulating properties and freedom from voids, low shrinkage, resistance to moisture, high mechanical strength and good adhesion to metals, plus the ability to be cast around conductors with complicated shapes in a simple operation. It was therefore possible to make insulators of a complex shape, with or without embedded conductors, in one simple casting operation.

Designers soon realised the economic advantages that could be gained by using epoxy insulation:

(a) Equipment could be upgraded.

(b) Equipment could be redesigned more cheaply as several functions in the original design could often be combined in one cast item. Frequently, although the epoxy casting was more expensive than the item it was replacing, the general improvement in design led to overall savings.
(c) Equipment could be reduced in size, with consequent savings in site area and building size of distribution stations.
(d) The complicated labour-intensive assembly of metal and insulating materials could be replaced by a simple casting process.

Over the past 25 years, epoxy resins have therefore become fully incorporated into switchgear in the 5–30 kV distribution range by the electrical industry in many countries, and there have been a number of useful reviews on the subject[1–5]. Below 5 kV, the high level of performance provided by epoxy resins is not usually required, and the resins have not been used above 30 kV because most stations operating at these voltages are usually sited outdoors, where conventional epoxy insulation has been considered unsuitable.

A typical example of the benefit derived from the use of cast epoxy insulation comes from Japan. In Japanese cities and towns, there is an acute lack of sites for urban substations, which has placed great emphasis on the need to design extremely compact switchgear. A range of switchgear rated up to 24 kV has therefore been produced by Mitsubishi Electric[6], using epoxy resin insulation wherever possible, which has enabled the overall size of the equipment to be reduced to very much less than that of an air-insulated assembly. Many outstanding examples of the European use of resin insulation were shown at the Hannover Fair in West Germany, in 1971[7], when a number of distribution transformers were exhibited which had primary and secondary windings encapsulated separately. This arrangement was very compact and also obviated the need for any special enclosure and for periodic checking of the cooling liquid level.

In general, cast resin continues to make a great impact on switchgear design and construction, enabling robust, compact equipment of pleasing appearance that is highly competitive in world markets to be produced. In addition to the cast resin for transformers and bushings, significant amounts of epoxy resin are used as laminates with glass-fibre to produce switchgear components such as mechanical switch linkages, arc quenching chambers and especially for the large, mechanically strong tubes used in air-blast circuit-breakers. Initially, cast epoxy resin was incorporated in switchgear operating in the 11–15 kV range, chiefly in order to gain service experience before proceeding to higher voltages and also because of the difficulty of incorporating stress control layers within the material. Thus certain air-break circuit-breaker units rated at 11 kV for indoor use make extensive use of cast epoxy resin for the insulation of bent connectors and their terminals. Orifice insulators are also cast integrally with the insulation of the connector and, in addition, the plug insulators have resin insulation.

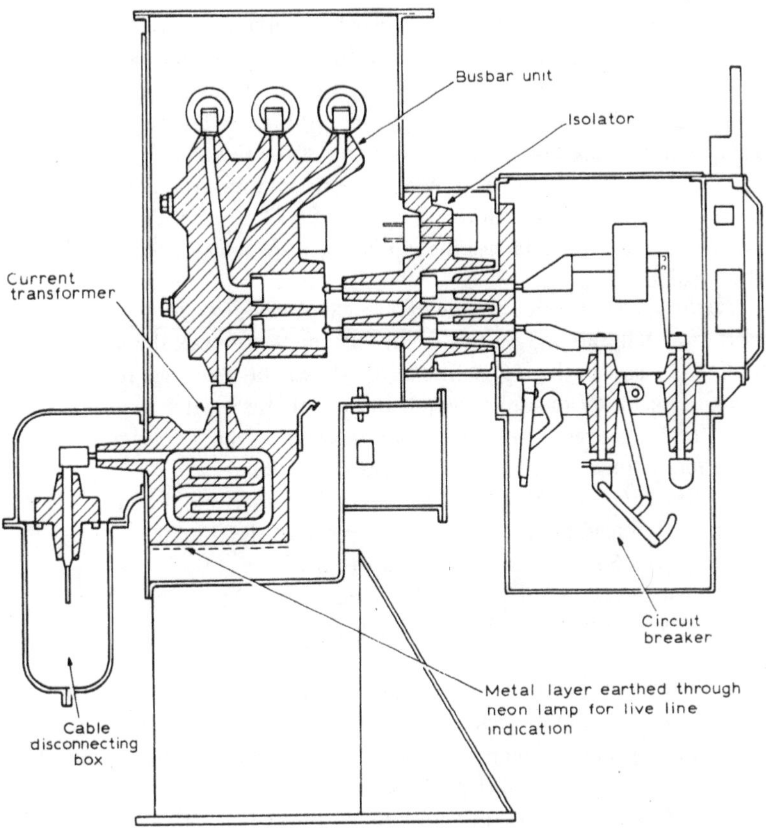

Figure 8.1. A 6.6 kV, 400 A, 75 mVA indoor oil-break, flame-proof circuit breaker unit with the cast resin-insulated components shaded (After Rothwell[1])

An oil-break unit also rated at 11 kV was mentioned by Rothwell[1] as incorporating a large amount of cast resin insulation for bushes and also for current and voltage transformers. *Figure 8.1* shows an indoor 6.6 kV, 400 A, 75 mVA oil-break, flame-proof unit with the cast resin-insulated components shaded to indicate the extent to which resin insulation has been used. The use of cast epoxy resin in this design has resulted in a very compact arrangement, which would have been almost impossible to achieve before the introduction of epoxy resin. In order to avoid exposing live busbars or cable connections to explosive atmospheres when the circuit-breaker is withdrawn, an isolating connector block is interposed between the circuit-breaker and fixed portion terminals. This single casting incorporates six high-voltage and twelve low-voltage connectors. The casting and circular flanges are machined to close limits so as to ensure a flame-proof path where they fit into the cylindrical metal casing. The plug contacts in the circuit-breaker are cast as a single unit, as are the orifice and busbars of the fixed portion. In *Figure 8.1*, the isolator is shown in the first withdrawal position in which the circuits within the flame-proof enclosure are disconnected. Further withdrawal detaches the switch from the isolator block and gives visible isolation.

Figure 8.2 shows an example of a single-piece epoxy resin casting used in switchgear, and illustrates the extremely complex shapes that can be cast in the resins. The figure shows the main cast resin block of an 11 kV ring-main switch and tee-off fuse-switch unit, which replaces over 20 separate insulating parts of a previous design. The overall dimensions of the block are approximately 870 x 300 x 550 mm. The use of a resin insulator has simplified assembly, reduced production time and reduced the overall volume of the unit to 60% of the design it supersedes[4].

The use of resin insulation for bushings in units operating above 15 kV was greatly assisted by two developments:

(1) The introduction of an insulation with a void-free high electric strength and low dielectric loss made from crepe paper impregnated with epoxy resin[8,9]. This paper, which is in tape form, improves still further the excellent mechanical and electrical properties of epoxy resins and further minimises shrinkage during cure. It also makes possible the insertion of metal foil layers for the control of electrical stress distribution

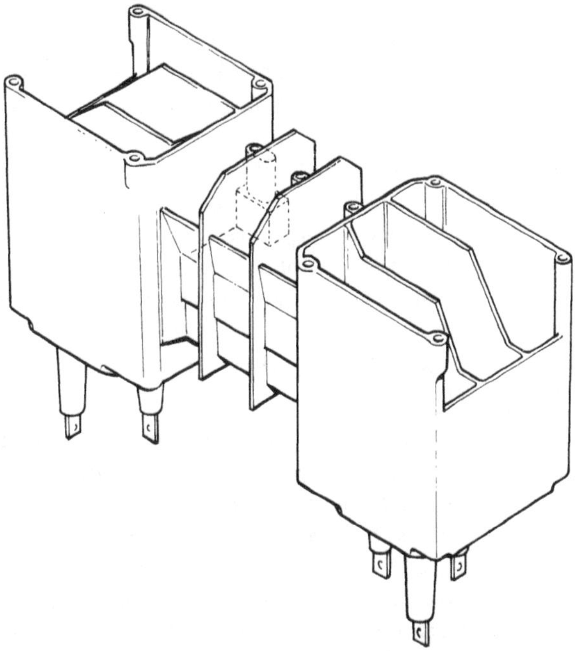

Figure 8.2. Cast epoxy resin insulator from an 11 kV ring-main unit (After Armstrong and Rutherford[4])

in the insulation, and enables a reduction to be made in the overall dimensions of the insulator, which can then be used at higher voltages than would be practicable with bulk insulation. This process also enables fully stress-graded insulation to be applied to conductors of uniform or non-uniform cross-section, which can be straight, angled or even toroidal. This is an advantage over bakelised paper or oil-impregnated paper bushings, as these can only be made in cylindrical form. This is exemplified in the case of a 33 kV outdoor circuit breaker[10], in which all three phases of the switch are contained in one oil tank. Here the use of angled bushings allows the appropriate minimum air spacing of terminals outside of the tank to be observed while at the same time allowing the use of a tank of minimum dimensions, thereby reducing the oil content and the weight of the equipment.

(2) The introduction of a corrugated paper tape impregnated with a silica-filled epoxy resin system[11]. This is incorporated into a casting as a skeleton structure of corrugated paper, and alternate layers of metal foil for stress control can also be inserted. The finished insulation is identical in outward appearance with conventional cast epoxy resins and has the advantage that suitably shaped creepage paths can be incorporated in a single casting process. This insulation is also a material of high electrical strength and low dielectric loss, which is resistant to moisture and completely gas-tight. It can be applied to a conductor of any shape and is ideally suited for bushings where grooves must be present in order to give a long creepage path.

Wall bushings (132 kV) containing insulation of this type have been used successfully in service since the early 1960s[2] and 380 kV terminal bushings weighing 1 ton have been produced and tested under switching conditions. There are now a large number of switchgear units, such as 33 kV outdoor oil circuit-breakers and 400 kV air-blast circuit-breakers which have been in satisfactory operation for some years and which incorporate large amounts of cast resin insulation.

Future trends in epoxy insulation for switchgear

Undoubtedly, this use of epoxy resin will continue to grow at a rate closely related to the growth of installed generating capacity in each country. In Western Europe, electricity production over the period 1950–70 grew at the rate of 7.7 per cent p.a., while installed capacity grew at the rate of 7.3 per cent p.a. However, it is not likely that these growth rates will continue to 1980, although the consumption of epoxy resins should increase during the 1970s, because of the possibility that: (a) epoxy systems will become acceptable for outdoor use, which will open up a further market (see Chapter 9) and (b) new designs will continue to favour the use of epoxy resins.

The trend towards liquid epoxy resins will increase, because of their easier handling characteristics, and anhydrides will continue to be the preferred curing agents. The low-viscosity weather-resistant epoxy resins may also begin to be used for indoor applications.

In the actual techniques of application, there will be continued effort given to increasing throughput and decreasing the labour content. Vacuum screw feeding direct to moulds could replace the current batch mixing, or a moulding could replace the casting process at least for small items of, say, a maximum weight of 1 kg. A further step might be the widened use of epoxy powder coatings to provide an even, insulating coating over complex-shaped busbars suitable for use in high-voltage equipment.

Epoxy–glass laminates in switchgear

Increasing amounts of epoxy–glass-fibre laminates are now being used in switchgear, the increasing size and rating of generating and transmission equipment forcing the industry to find insulating materials with higher mechanical capabilities. An attractive feature of these laminates is their very high strength to weight ratio and, in addition, they have good electrical properties and physical inertness in electrical cooling and insulation media and they retain these characteristics over a wide temperature range.

In many of the laminates used, unidirectional oriented glass-fibres are employed because of the ability to obtain a wide range of mechanical characteristics in the laminate by varying the fibre orientation in different layers of reinforcement. The unidirectional laminates have been used successfully for the fabrication of switchgear linkages and lift plates. A further use has been in the manufacture of insulated control shafts of transformer tap changer selector switches. In this use, it is an essential part of the design that there is an exactly syncronous movement of each of the phases, i.e. there must be no significant twisting or dimensional lag between them during operation. Large numbers of these shafts have been in operation for some time with completely satisfactory results.

In both of these units, epoxy–glass laminates have also been used, either for the arc control device housing or for saddles from which the moving contacts are suspended.

Large amounts of epoxy–glass laminates are used in switchgear for air-blast tubes and arc chambers, together with numerous linkages, connecting rods and levers used for moving the copper contacts. In modern heavy-duty air-blast circuit-breakers, the arc is attenuated and extinguished simply by blowing it out by means of an extremely powerful blast of air at a very high pressure. This procedure produces virtually a shock wave and the tube along which the air blast is directed must be very strong. Such tubes must also be insulators in addition to possessing high mechanical strength, especially tensile strength, under shock loading conditions.

Porcelain–phenolic resin-impregnated paper tubes are not strong enough to withstand repeated mechanical shocks of this magnitude, whereas epoxy–glass laminated tubes are able to meet these severe design criteria. The size of these tubes obviously depends on the mechanical design of the circuit-breaker and its power rating. For example, one design of circuit-breaker suitable for 275 kV distribution systems requires cylindrical tubes 2 m long and 15 cm in diameter. Some 400 kV breakers require tubes 20 ft (6 m) long and 12 in (28 cm) in diameter.

Most epoxy resin systems possess the necessary insulating properties for the very high voltages found in service, but when used to form a laminated tube with glass-fibre it is essential that the walls of the tube should be void free. Under high-voltage stresses, discharge effects can cause rapid breakdown of the laminate if it is not void free, although the resin system is satisfactory. A further requirement is that the tubes should possess a high tensile strength together with good tensile elongation. In practice, it is necessary to compromise and to maximise these two properties as far as possible, because they tend to be mutually incompatible. It has been found that a resin system with a tensile strength of 9000–15 000 p.s.i. (63–105 MN/m^2) will produce excellent laminated tubes provided that the tensile elongation at break is at least 5 per cent.

The tubes can be made by either vacuum impregnation or filament winding. The latter method will produce stronger tubes provided that a suitable winding pattern can be adopted. However, the complete exclusion of air bubbles or voids is extremely difficult during this wet winding process and, for this reason, the preferred technique has become that of vacuum impregnation as it produces completely homogeneous laminates of high electrical strength.

TRANSFORMERS

Transformers are used whenever an alternating current is used. They are available in a wide range of sizes, from very small, which handle millivolts and milliamps, up to very large, which operate in the kilovolt and kiloamp range. This section is concerned with those transformers that operate in the kilovolt range and above, which are used for high-voltage electricity distribution. The very small units are considered in Chapter 10.

A wide variety of materials are used as insulators in transformers, including air, wax, bitumen, oil, chlorinated diphenyls and thermosetting resins such as phenolics, polyesters, and epoxies; the choice of the insulant depends not only on current and voltage requirements but also on physical size, temperature classification, environmental conditions and economics. In general, epoxy resins are used when:

(1) Severe environmental conditions are to be encountered such as high temperatures and humidity.
(2) Mechanical shocks due to sudden changes in electrical load occur, for instance in current-monitoring transformers. The use of cast epoxy insulation in transformers of this type enables the equipment to withstand the overload currents that occur, and the consequent very severe mechanical and thermal stressing that results from this overloading[12].
(3) Design improvements and cost economies can be made. The use of epoxy resin insulation can allow transformers to be redesigned to achieve (a) size and weight reductions, often up to 50 per cent of the original, (b) more dependable service under higher temperatures, up to 70 °C, (c) stronger construction by bonding core, coil and terminals together into one solid unit and (d) a smooth, hard exterior which improves appearance[13].

Many small supply transformers for use up to 5 kV contain epoxy and also polyester insulation in order to reduce size and weight, but the choice of insulant is determined primarily by the size of the unit and the economics of production. In Europe, most distribution transformers are totally immersed in oil, a technique that has been shown over the years to be both economical and reliable. In the U.S.A. smaller distribution transformers are often mounted on poles, where size and weight reduction are of great importance and epoxy insulation is often used.

In medium-voltage switchgear, dry voltage transformers in which epoxy insulation is used are now well established[3]. Primary windings of large wound transformers often have paper interlayer insulation, and the coils are usually impregnated with resin and then completely embedded in the resin; high-voltage bushings can then be cast integrally with the unit. In this type of transformer for voltages greater than 33 kV, a different technique is often used, whereby the high-voltage winding is impregnated and, while still wet with uncured resin, a filled epoxy is cast around the windings. Current transformers for use in switchgear at voltages up to 15 kV are also encapsulated in epoxy resin, either the primary or secondary windings being embedded in resin, depending on the design of the unit.

The very large transformers used for stepping up the voltage of the current generated in power stations for distribution over long distances are frequently of the oil-immersion type, although epoxy resins are being used increasingly. For example, in the U.S.A., power transformers in the 750 to 1000 kVA range use epoxy insulation.

In a similar way to their use in switchgear, epoxy resins have enabled new designs of transformers to be developed[14,15] that are lighter, smaller and more reliable than earlier designs and have improved protection from the environment. These characteristics are particularly favourable if the equipment is to be used underground in mines where there is a severe height limitation or in industrial and commercial situations, or even on ships, where the optimum use of space is very important.

For many years, the basic structure of transformers has remained fundamentally unchanged. The coils are wound using round or semi-round copper wire and insulated turn from turn with paper or enamel. Paper, press-board or other similar material is used to insulate between layers and separate coils. The entire structure is then vacuum impregnated with oil. A new design of distribution transformer involves the use of epoxy enamel-coated aluminium foil and total encapsulation of the wound foil in an epoxy formulation[16]. The combination of insulated foil conductor and total encapsulation has resulted in a transformer which makes more efficient use of the core window space. Moreover, high stresses between turns and layers, which are normally produced with round wire and layer insulation, have been eliminated and a more uniform impulse voltage stress produced. Overall, the units use less insulation than previously and do not require the complicated drying procedures associated with oil–paper systems, and hence overall cost savings can be made[17].

EPOXY CABLE JOINTING COMPOUNDS

Sealing compounds are normally used to ensure that joints in cables are effectively sealed against the ingress of water and protected from mechanical damage which could result from underground cables being struck by digging implements.

The traditional method of protecting cable joints is to place a cast iron box around the joint and to fill it with hot bitumen, which, on cooling, forms a solid protective casing. Lead-sheathed cables can also be sealed by plumbing the sheath. These methods were acceptable when cables contained oiled paper insulation within a lead sheath. However, the introduction of plastic and rubber insulation and sheaths in the early 1960s meant that joints could not be plumbed, and adequate adhesion of bitumen to thermoplastics was difficult to obtain. Attention was therefore turned to the cold pouring synthetic resin sealing compounds, which, typified by epoxy compounds, have the following advantages over bitumen:

(1) They are more compact and eliminate the need for operatives to carry bulky iron boxes or lead sleeves. No vehicle is needed to carry the material between jobs.
(2) No heating is required, in contrast to the need to heat the conventional bitumen jointing compounds. Moreover, careful control of the bitumen temperature is required as excessively hot bitumen can damage cable insulation and cause plasticiser migration in PVC sheaths, which can lead to poor adhesion and embrittlement of the PVC.
(3) If the epoxy is used as a kit with pre-weighed materials, there is no need for stirring or measuring equipment.
(4) The cured epoxy system has low shrinkage during cure and does not creep or flow during service.

All of these reasons encouraged the further examination of resinous compounds, and synthetic resin jointing compounds are now widely used in telecommunications cables in many countries. Further, increasing interest is being shown in their use for sealing electricity distribution cable joints, although the problems involved in acceptance for this use have taken longer to overcome.

REFERENCES

1. ROTHWELL, K., *Elect. Rev., Lond.*, **177**, No. 3, 89 (1965)

2. LEGG, D. and ROTHWELL, K., *High Voltage Applications of Epoxide Resins*, CIBA, 63 (1967)
3. BARKER, H. and FLACK, T., *9th IEEE Insulation Conf.* (1969)
4. ARMSTRONG, T. V. and RUTHERFORD, R., *Elect. Times*, **155**, No. 31, 36 (1969)
5. GARRARD, C. J. O., *Proc. IEEE*, **116**, No. 6, 1031 (1969)
6. *Elect. Rev., Lond.*, **189**, No. 5, 161 (1971)
7. *Elect. Rev., Lond.*, **188**, No. 20, 653 (1971)
8. IMHOF, A., *CIGRE Rep.*, 119 (1962)
9. MOSER-GLASER, AG, *Brit. Pat.* 886, 290
10. *Araldite in Stress Graded Insulators*, CIBA Technical Literature
11. REYROLLE LTD., *Brit. Pat.* 967, 853
12. MERCIER, C. E. *Allis Chalmers elèct. Rev.* **21**, No. 2, 24 (1956)
13. WAHLGREN, W. W., *Insulation, Lake Forest*, **4**, No. 6, 14 (1958)
14. MCWHIRTER, J. H., FOSTER, N. C., BERG, D. and HOFMANN, C. F., *Westinghouse Engr.*, **23**, No. 1, 13 (1963)
15. ELY, E. R., KURZ, R. A., PALUMBO, A. J. and ZICKAR, F. R., *7th IEEE Insulation Conf.* (1967), Paper No. 32, C79-72
16. *Insulation, Lake Forest*, **10**, No. 8, 27 (1964)
17. HOFMANN, C. F., *SPE J1*, **25**, 74 (1969)

PLATE 4
Refrigerators painted with epoxy resin powder coating (Courtesy of Shell)

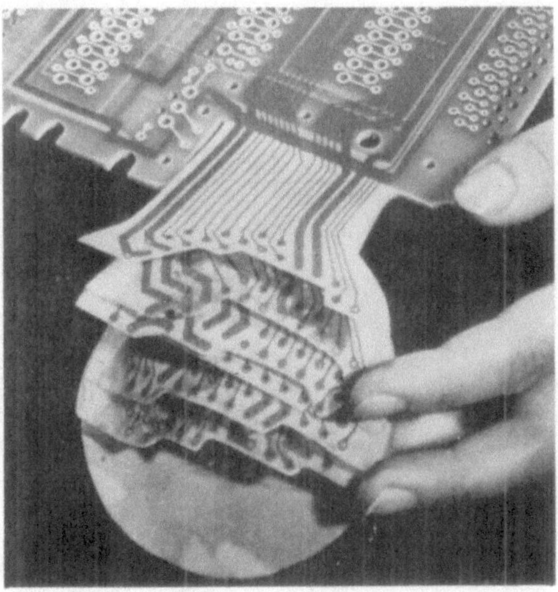

PLATE 5
An epoxy resin based thirteen-layer circuit board incorporating twelve external flexible connections, i.e. 'flying tails'. Manufactured for Ferranti Ltd. by Technograph & Telegraph Ltd. (Courtesy of CIBA-Geigy)

PLATE 6
Epoxy resin based rotor slot liners by C.A. Parsons & Co. Ltd., Heaton Works, Newcastle-upon-Tyne, 6 (Courtesy of CIBA-Geigy)

PLATE 7
Epoxy resin in transformers for welding equipment by British Federal Welder & Machine Co. Ltd. (Courtesy of CIBA-Geigy)

9
Epoxy Resin Insulation for Outdoor Use

The previous chapter described how cast or laminated epoxy systems have replaced to a large extent the traditional materials porcelain, oil, and laminated paper in many items of equipment used indoors or in a protected situation. The main advantages in using epoxy insulation that have brought about this change have been a high strength to weight ratio, together with good electrical performance that enables equipment to be redesigned so as to obtain valuable weight and space savings. These advantages could probably be utilised in equipment that is employed outdoors, and much effort has been and is still being expended on the development of epoxy resin systems that are suitable for outdoor use.

All outdoor power distribution networks consist of the line conductors and their insulation and supports, together with transformer sub-stations in which the high voltage power is converted to a level suitable for distribution to smaller areas. These distribution points use many insulators and bushings, together with items of switchgear and ancillary equipment. Where the equipment needs to be protected it is enclosed in metal boxes and, in fact, some smaller sub-stations are totally enclosed and much cast epoxy resin insulation is used in this protected equipment for insulators, bushings and various switchgear components. Where electrical equipment is exposed to the atmosphere, the usual insulating materials are still ceramic or glass, and this includes line insulators as well as equipment in the sub-stations.

Up to the present, the main insulation for power lines has been provided by ceramic or glass insulators mounted either on wooden cross-arms supported by wooden poles for lower voltages or on steel cross-arms supported on steel towers for higher voltages (132 kV or more). Because of increased power requirements, the trend now is to use higher and higher transmission line voltages and, in the U.K., line voltages have increased from 132 to 275 and then to 400 kV. Plans are already in hand for 700 kV installations and no doubt 1000 kV systems are possible, provided that suitable support structures and insulation can be found. Higher voltages demand extended leakage paths between the line and the supporting tower, and therefore insulators have to be longer and larger in diameter. Inevitably, ceramic insulators increase in weight as their overall dimensions increase and their designs become more complicated and cumbersome. Moreover, the supporting tower would need to be increased in height and strength. All of these factors lead to large cost increases, the need for larger areas of land for the siting of the bigger towers and, equally important, the line of the towers might be an even greater eyesore than before in many situations.

In the light of these difficulties and particularly the inadequacy of traditional materials for high-voltage working, the electrical industry has had to reconsider the whole power line and distribution system. The structural strength needed by the support towers could easily be provided by glass or carbon fibre-reinforced epoxy resin members. Replacement of ceramic or glass insulators by other materials which would need to be lighter in weight and able to withstand severe electrical stressing in a difficult environment is less easy to achieve. However, it is this part of the transmission line system that is more likely to use large amounts of epoxy resins in the future than the replacement of traditional materials by laminates in the construction of supporting towers.

Many polymeric materials possess the electrical properties required by an insulating material and could function satisfactorily under severe atmospheric conditions without surface breakdown. Unfortunately, most of these materials do not possess the required mechanical strength. Epoxy resins are clearly materials that must be considered, as they possess good electrical properties coupled with high mechanical strength, especially when used in conjunction with glass-fibre. However, improvements must be made in the weathering characteristics and tracking properties of epoxy systems before these resins can be seriously considered for outdoor insulation.

In the late 1950s and early 1960s, the solid epoxy casting resin system was evaluated for outdoor use. The resin was used to make traditional ceramic shaped insulators, which were then exposed to electrical stressing. Discouraging results were obtained, a typical result being failure by tracking after exposure for 2½ years in a semi-rural environment at 1½ kV/in stress. Further experimental work confirmed that, especially under conditions of high electrical stress and severe atmospheric conditions, epoxy resins that possess an aromatic 'backbone', typically the DPP–ECH glycidyl ethers, do not give a satisfactory performance. The surface of castings of these resins breaks down by crazing under the influence of ultra-violet light and humidity, and these surface cracks readily trap dirt particles from polluted atmospheres. Under high electrical stress, surface tracking occurs along the path of these trapped dirt particles and, once this short-circuit path has been established, the resin carbonises along it when the current passes, setting up a permanent tracking path. Such a performance compares unfavourably with those of glass and ceramic insulators, which do not suffer from permanent surface breakdown under 'flash-over' conditions. When these insulators are covered with industrial pollution or marine salt, their performance can be restored by cleaning, and a service life of 20–30 years is typical.

Improvements in the performance of epoxy insulators were therefore sought, and the following approaches were tried:

(1) Redesign of the insulator to make the operating conditions less stringent for epoxy resins.
(2) Development of new types of epoxy resins with improved weathering and tracking resistance.
(3) Use of an epoxy core with tracking-resistant weather sheds.
(4) Coating a traditional glycidyl ether resin insulator with materials with superior weathering characteristics.
(5) Improvement of the tracking resistance of the glycidyl ether resins by the use of hydrated alumina as a filler.

Of these approaches, (1) in combination with (2) is probably the most likely to gain acceptance for outdoor insulator construction. The high strength to weight ratio of epoxy resins can make possible the production of cast or moulded shapes that would be impracticable with ceramic or glass. Insulators with thin sheds could be produced, and would allow an increased surface area to be given to the insulator. The leakage path of the insulator over a given overall length could therefore be increased, thus reducing the electrical stress per unit of length. While it might be possible to use the DPP–ECH epoxy resins in this application, the industry would clearly wish to have the additional safeguard against premature failure provided by an improved tracking-resistant resin.

Technique (3) has been developed to a practical scale in Italy, where small railway insulators are being made from a polyester extruded rod core with PTFE sheds sealed to the core with a flexible epoxy adhesive.

However, difficulties are likely to occur if adhesion between the core and the PTFE sheds is lost and moisture is allowed to penetrate into the joint. The design of an insulator of this type, in which the stress-carrying central core has sheds of a dissimilar material bonded to it, has an inherent weakness at this point of juncture between the two materials.

The use of weather-resistant coatings applied to epoxy insulators, method (4), remains a possible solution to the problem, despite certain adhesion failures of coatings after 3–4 years' weathering.

The use of an anti-tracking additive in epoxy systems, method (5), was considered at an early stage and the use of trihydrated alumina for this purpose was patented. This method still shows promise despite the reduction in mechanical properties brought about by this filler compared with silica-filled systems.

DEVELOPMENT OF WEATHER-RESISTANT EPOXY RESINS AND ADDITIVES

American work established the value of trihydrated alumina as an anti-tracking additive in the late 1950s, and at about the same time, theoretical studies were being carried out in Europe[1]. Subsequently practical trials confirmed the theory that tracking resistance is related to the chemical nature of an organic polymer and more particularly indicated that completely saturated compounds are to be preferred. Resin systems of this type not only possess a greatly reduced tendency to form carbon on pyrolysis, probably owing to the absence of aromatic rings, but also show an improved resistance to ultra-violet light with a consequent reduction in crazing and surface breakdown. Retention of dirt on the insulator surface is therefore decreased.

Two main types of non-tracking epoxy resins have now been developed:

(a) cycloaliphatic resins containing a 7-oxybicycloheptane ring, and
(b) glycidyl ester resins derived from cycloaliphatic acids such as tetrahydro or hexahydrophthalic acids.

Both types of resin must be cured with an aliphatic acid anhydride in order to realise their full anti-tracking potential. In general, the cycloaliphatic resins of type (a) tend to produce hard, inflexible cured products with high heat deflection temperatures, while the glycidyl esters give lower heat deflection temperatures and mechanical properties of the same order as those of the glycidyl ether DPP–ECH epoxy resins[2].

USES FOR OUTDOOR EPOXY INSULATION

Adopting one or more of these various approaches, the possible use of epoxy insulation is being actively considered for:

(a) Insulators in high-voltage lines.
(b) Insulators in low-voltage lines.
(c) Insulated cross-arms for novel tower design.
(d) Insulators in railway overhead lines.
(e) Redesign of equipment, e.g. transformers.

High-voltage line insulators

Insulators on high-voltage transmission lines, such as the 400 kV lines, are constructed from porcelain or glass, but epoxy resin insulators would have a considerable advantage in weight if they were acceptable on technical and cost grounds. The advantages of epoxy over porcelain insulators are, in summary[3].

(1) Lower weight. Cast epoxy resin insulators are merely one-third of the weight of porcelain insulators.
(2) Greater accuracy in dimensional tolerances. Dimensional variations in porcelain insulators can cause installation difficulties.
(3) Superior mechanical properties. The tensile, flexural and impact strengths of epoxy resins are vastly superior to those of porcelain. In addition, unlike porcelain insulators, epoxy resin insulators are virtually free from internal stresses and are much less sensitive to sudden variations in temperature.
(4) Ease of attachment of metal inserts. Unlike porcelain, epoxy resins permit the direct inclusion of metal inserts into castings. In addition, the dimensions and shapes of metal inserts may be designed so as to avoid points of high electrical stress at which breakdown may be initiated.
(5) Superior dielectric properties. The dielectric properties of a cast epoxy resin system are compared with those of porcelain in *Table 9.1*.

Table 9.1

DIELECTRIC PROPERTIES OF EPOXY RESIN AND PORCELAIN

Property	*Epoxy resin system*	*Porcelain*
Dielectric constant	3.7–4.2	5.0–6.5
Power factor	0.009–0.014	0.02–0.03
Volume resistivity	15 $\log_{10}$ ohm cm	approx. 12 $\log_{10}$ ohm cm

The limitation to the use of epoxy resins for insulator manufacture, namely their poor tracking resistance under polluted conditions, has already been mentioned, and some of the approaches that are being explored to overcome this problem described.

Trials have been in progress in Europe and the U.S.A. for many years in an attempt to evaluate the various possible ways of producing satisfactory epoxy insulators. The results of the trials so far suggest that in conditions which are not excessively severe, i.e. absence of high atmospheric pollution and at medium voltages, epoxy resin suspension and tension insulators would perform satisfactorily for many years, although they would probably not have as long a life as porcelain insulators. Both types of insulators are loaded in tension and are able to take advantage of the great strength that epoxy–glass laminates exhibit in this mode of stressing.

However, the test programme showed that cracking of the resin occurred in some composites at a far lower tensile stress than had been anticipated, which was possibly associated with the presence of glass near the surface of the composite. Poor adhesion of resin to glass in this situation would allow moisture penetration to occur, which could initiate further adhesion failure. Accordingly, the design of epoxy insulators has tended towards high-strength glass laminate cores made with a flexible resin system together with flexible weather sheds either cast on to the core or pre-cast and subsequently bonded to the core. In this way, the shed material can be made free from glass and given optimum electrical performance without regard to mechanical strength. A typical example of such an insulator is a spun-cast, rigid alicyclic resin system filled with aluminium trihydrate and silica for the sheds, an extruded unidirectional resin–glass rod as the load-bearing member and a highly flexible unfilled system to couple the shed to the load-bearing rod[4,5]. A further possibility would be to use a filament-wound epoxy–glass tube as the central strength-bearing core rather than a pultruded rod. Both designs are acceptable propositions provided that the mechanical stresses are suitably distributed and the strength of the glass is fully utilised.

A further design for insulators for 132 kV tests consisted of a glass-fibre loop between two metal bushes, shrouded by a core of resin at least 20 mm thick (see *Figure 9.1*). The insulator weighed 60 kg and had a working load of 104 kN. However, there were indications that weak points were likely to develop around the end bobbins due to the development of stress on cure. Nevertheless, the results obtained were

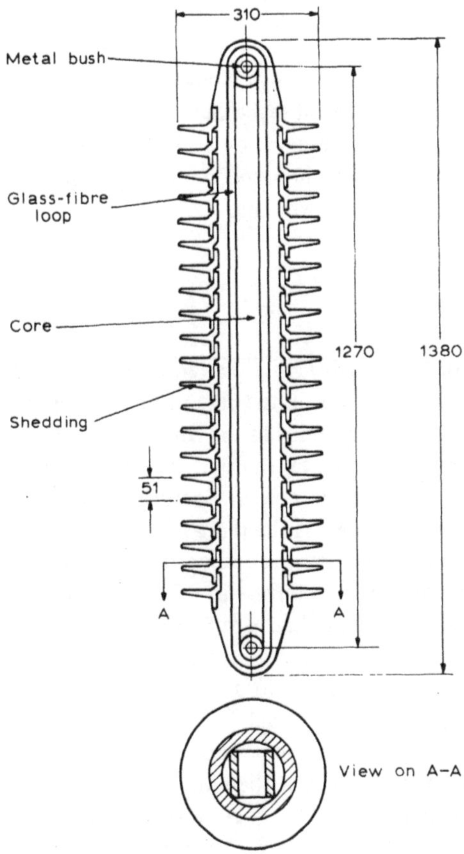

Figure 9.1. Experimental resin-bonded glass-fibre insulator for 132 kV[6]*:* Dimensions are given in millimetres

useful from the point of view of future large-scale applications of epoxy resin-based insulators.

The economics of the use of epoxy insulators of this last type have also been considered. A set of 2 x 3 (series) of these 132 kV units would be about 2½ times as expensive as a standard 4 x 22 unit glass or porcelain set for a 400 kV quad line[6]. The cost of the standard set in 1969 was about £195 and it weighed 1400 kg, whereas the six resin units would have cost about £486 using silica-filled anti-tracking aliphatic resins and would have weighed about 360 kg. Therefore, in order to be competitive, the resin units must be further reduced either in weight or price. It is possible that both of these factors can be reduced if the insulator can be built round a hollow laminated cylinder, but clearly more design work is required. A further consideration is that this cost comparison is between a well developed ceramic process and a very simple epoxy construction technique. The unfavourable result for the epoxy system could be reversed by an enlightened approach on manufacturing, using modern high-speed techniques such as moulding, pressure casting and filament winding.

Low-voltage line insulators

The advantages of the weight saving that could be achieved by using epoxy insulators were obvious in the case of high-voltage lines, but these savings are not as important in the case of low-voltage line insulators. However, there are far more of the small insulators used than the large insulators and in terms of the potential market for epoxy resins the small insulator field is the more attractive.

What incentive is there for using epoxy insulators in low-voltage applications? One reason for their use in the U.S.A. has been their resistance to breakage and vandalism, but this is not regarded as important in Europe. Reported sales of these insulators in the U.S.A. also suggest that the U.S. market does not regard breakage resistance to be as important as was once believed.

If it proves possible to produce pollution-resistant epoxy insulators, there might be an incentive to use them in polluted environments, but for widespread acceptance and use they must be available at a similar price to those of porcelain and glass. Comparison of raw material costs shows that an epoxy system is about ten times more expensive than porcelain, but the selling price of epoxy insulators is only two to three times greater than that of porcelain insulators. The production process for ceramic insulators is well established (probably fully developed) and involves high heat loss and wastage rates, and it is therefore unlikely that a significant reduction in production costs for ceramics can be made. In contrast, the present methods of making epoxy insulators are crude and undeveloped, which suggests that the probable improvements in the epoxy process are likely to lead to a price for epoxy insulators which approaches that of the ceramic insulators. Provided that technical acceptance can be obtained, this situation would lead to a very large potential market for low-voltage epoxy insulators.

Insulated cross-arms

In the U.K. there are three types of overhead line structures:

(1) Straight line structures for voltages up to 66 kV, where the line conductors are usually above the cross-arm and supported by pin-type insulators.

(2) Straight line structures for 66 kV up to the highest voltages where the conductors are supported from cross-arms, usually by porcelain or glass insulators in series, and these are able to swing freely.

(3) Tension insulators are used as angle and terminal structures and are in series with the line conductors, being connected directly to the cross-arm or another part of the power structure.

If the line conductors in types (1) and (2) could be connected directly to the cross-arms, it would be possible to eliminate the insulators. In order to achieve this, a material for the cross-arms would be required that had outstanding electrical properties and acceptable mechanical properties. In addition to the elimination of insulators, another advantage would be the simpler construction of the wooden type (1) towers. Type (2) steel towers could be reduced in height and weight and require a smaller area on the ground. *Figure 9.2* shows a comparison of a conventional tower and a tower with insulated cross-arms designed to carry 400 kV line voltages[7]. The potential height reduction of nearly 30 ft is immediately apparent.

In addition to their use in the construction of new lines, resin-insulated structural members could also be used to uprate existing lines to higher voltages at 25–40 per cent of the cost of installing a new line[7].

In most countries there exists a need to meet the ever-increasing demand for electrical power, and uprating lines to higher voltages is therefore an important alternative to dismantling existing structures and replacing them with new ones. Trials have been under way in the U.K. since 1966 on uprating 33 and 66 kV lines to 132 kV and are proceeding satisfactorily (1974). One of the first commercial installations of insulated cross-arms[8] occurred where a 33 kV line was converted to 66 kV, and more recently a new design of 132 kV line has entered development, using epoxy post type insulators[9].

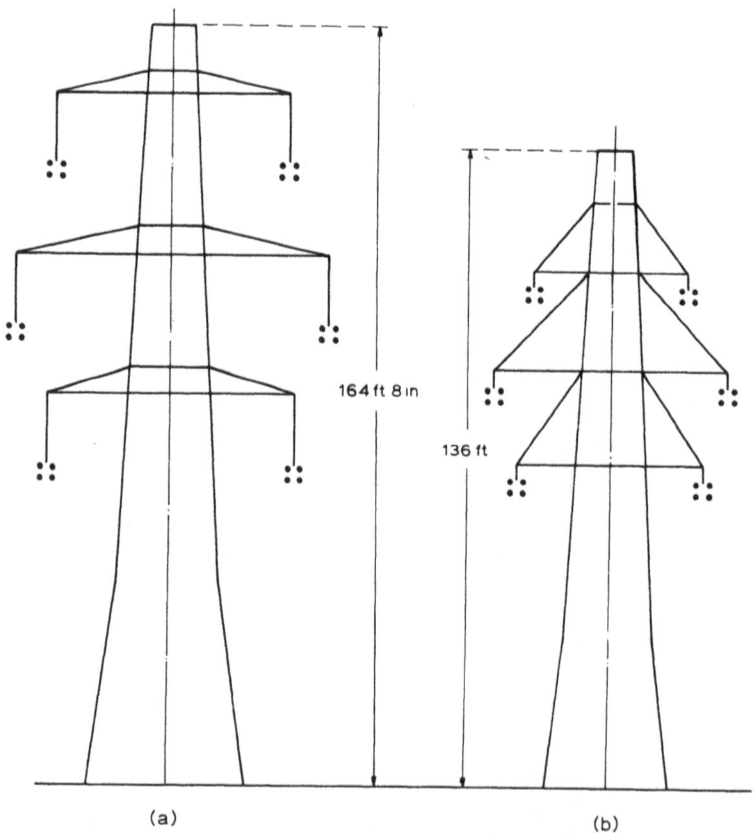

Figure 9.2. Comparison of (a) *a conventional 400 kV tower (C.E.G.B. 400 kV quad 0.4 tower) with* (b) *a tower with insulating cross-arms* (After Day *et al*[7])

Insulators in railway overhead lines

In Italy and Eastern Europe, a considerable number of resin insulators have been installed for railway power transmission lines. These have been constructed from a polyester–glass pultruded core with moulded PTFE sheds which are bonded to the core with a flexible epoxy adhesive. In general, these insulators have performed satisfactorily although some failures have occurred owing to loss of adhesion between the core and the sheds.

The significant point about this development is that by using an efficient manufacturing process, it has been found possible to sell a resin insulator at less than twice the price of an equivalent ceramic unit. Clearly, the advantages of the resin insulators, such as light weight and freedom from breakages, have also enabled the manufacturer to obtain a price premium over ceramic insulators. However, for resin insulators to capture a large part of this market, it is probable that they will need to be offered at about the same price as ceramics.

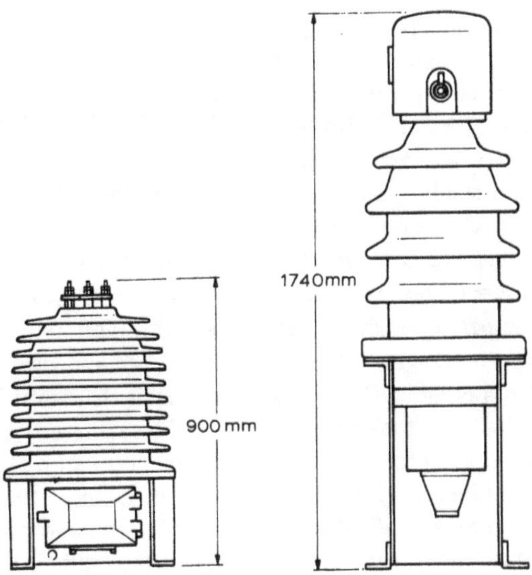

(a) (b)

Figure 9.3. Comparison in size of (a) *a 72.5 kV cycloaliphatic current–voltage transformer developed by Moser-Glaser and Co. and* (b) *a transformer for the same voltage but with conventional oil insulation* (Courtesy CIBA-Geigy Ltd.[10])

Redesign of equipment

The outdoor applications of resins include their use in current and voltage transformers. A number of electrical companies in countries such as Finland, France, West Germany and Switzerland have successfully developed and marketed items of equipment such as post insulators, bushings and current and voltage transformers up to 69 kV and also a combined current and voltage transformer operating at a maximum voltage of 72.5 kV. *Figure 9.3* shows clearly the reduction in size that can be achieved with the latter equipment when cast epoxy resin is used in place of conventional oil insulation[10].

Porcelain insulators in the 138–345 kV range for power circuit-breakers have become very large and consequently very difficult and expensive to build. An epoxy resin outdoor casting could reduce the weight of the insulator by 40 per cent compared with the equivalent porcelain design, and be much quicker to produce[11].

OTHER MATERIALS FOR INSULATORS

Polymers other than epoxy resins have also been considered for the construction of complete insulators and cross-arms, but none appears to offer a better combination of mechanical strength, weather resistance and electrical performance. However, PTFE and silicone rubber appear to be satisfactory weather shed materials and could be used on a large scale as weather sheds in conjunction with polyester–glass or epoxy–glass strength cores, but the problem of obtaining and maintaining adequate adhesion between the cores and the PTFE sheds remains.

Manufacturers of ceramic and glass insulators have also been seeking to improve the performance of their products. Undoubtedly, ceramics or glass of higher strength or lower weight will become available, but it is unlikely that a combination of these properties that is sufficiently attractive to challenge the epoxy systems will be achieved. An important development has been that of stabilised ceramic insulators, involving the use of a semiconducting glaze on the surface of the insulator. The glaze, which contains small amounts of tin oxide and antimony oxide, dissipates about 7.5 W to the atmosphere, which in still air is sufficient to warm the insulator to 10 °C above the ambient temperature. Under normal conditions of natural weathering, this is sufficient to keep the insulator dry, and the insulator would be effectively pollution proof.

This use of a semiconducting surface on an insulator could also be considered for epoxy as well as ceramic units, and would have the added advantage of minimising low-temperature embrittlement of the system and also preventing surface pollution.

THE FUTURE

As far as overhead transmission line systems are concerned, considerable cost savings could be achieved by the use of epoxy resins in insulators and cross-arms. So far, field and other assessments have shown that more work needs to be done in the design of both insulators and cross-arms so as to ensure that the properties of epoxy resins are used to the maximum advantage. Of the resin systems themselves, present experience of their use is not sufficient to justify their widespread use in all situations, but there are certainly many opportunities where overhead line equipment fabricated from epoxy resins, such as tension insulators and cross-arms being used to upgrade line voltages, would give an acceptable working life.

In order to become fully accepted, epoxy systems must offer clear improvements over the traditional materials at a similar price. This is a challenging requirement for the resin manufacturers, but a large potential market exists for the manufacturer who can gain this acceptance. Little help in the development of resin insulation can be expected from the national authorities responsible for generating and transmitting power who, in general, regard the responsibility for developing new insulation materials as resting with the equipment manufacturers. It is evident that for some of the problems discussed in this chapter to be solved, a wide range of skills and experience is required, for example in resin technology, the use of resin reinforcement and in electrical, mechanical and production engineering. This will best be obtained by the collaboration of interested companies, each of which can contribute expertise in each of the appropriate fields, and such associations of firms have already begun to be formed.

REFERENCES

1. PARR, D. J. and SCARISBRICK, R. M., *Proc. Instn elect. Engrs,* **112**, No. 8, 1625 (1965)

2. *Epikote Resin Systems for Track Resistant Castings,* Shell International Chemical Company
3. *Araldite Epoxy Resins for Outdoor Insulation,* CIBA-Geigy Ltd.
4. MCINTYRE, J. M. *Proc. Instn elect. Engrs.* **116**, No. 12, 2040 (1969)
5. LOOMS, J. S. T., *Elect. Rev., Lond.,* **188**, No. **4**, 117 (1971)
6. STANNETT, A. W., LAMBETH, P. J., PARR, D. J., SCARISBRICK, R. M., WILSON, A. and KINGSTON, R. G., *Proc. Instn elect. Engrs,* **116**, No. 2, 261 (1969)
7. DAY, P., NICHOLSON, J. A. and PROCTOR, F. H., *High Voltage Applications of Epoxide Resins,* CIBA, 115 (1967)
8. *Elect. Times,* **157**, No. 11, 23 (1970)
9. PUGH, G. B., *Elect. Rev., Lond.,* **193**, No. 18, 597 (1973)
10. *Outdoor Electrical Insulation with Araldite Resins,* CIBA-Geigy Ltd.
11. HOFMANN, C. F., *SPE Jl,* **25**, 74 (1969)

10 Electronic Applications; Printed Circuits

ELECTRONIC APPLICATIONS

Electrical and electronic equipment needs to be protected against the environment, and also often needs mechanical support. For many years, protection from moisture, dirt and physical abuse was provided by hermetically sealed metal cases, and this technique is still used for certain items, but for many components, adequate protection is now obtained by encapsulation in a resin, and at a lower cost[1].

Manfield[2] gave some indication of the comparative costs of protecting a semiconductor device by hermetic sealing and by encapsulation in resin. He stated that resin materials are much cheaper than glass to metal seals and metal cases, labour costs are about equal, but overheads are much lower when resins are used. Even over a long period of time, when large numbers of items have been made, the resin method is cheaper than hermetic sealing.

Embedment in resin not only protects the components but also stabilises electrical properties and prevents leakage, protects against excessive mechanical pressure and eases handling. The need for equipment with improved reliability and protected in this way was first expressed by the military authorities, for use in aircraft radar and rocket guidance and telemetry systems. Nowadays, stringent reliability demands are also made for components to be used in aircraft, computers and consumer durables, such as washing machines. The necessity to provide a protective barrier for semiconductor devices is apparent when an integrated circuit is considered. This circuit may be as complex as a radio receiver and yet be contained in a thin film measuring 4.5 mm square and having perhaps ten or more leads attached, each about 0.025 mm (0.001 in) in diameter. Such a device is prepared under conditions of high purity and must be protected in order to preserve its characteristics.

The range of components protected by resins is very large and includes capacitors, resistors, chokes, relays and printed circuit and electronic circuit modules as well as semiconductor devices such as transistors, diodes, thyristors, thin film resistors, integrated circuits and many others [3-6]. For over 20 years, certain classes of these components (those which Manfield[2] termed 'passive'), such as capacitors, resistors and transformers and a number of others, have been encapsulated in epoxy resins. The use of resin encapsulation for the other classes of components (Manfield's 'active' group), which are basically the semiconductor devices, has not yet become fully established. While the advantages of resin encapsulation, such as lower costs and better heat dissipation, are acknowledged, the long-term reliability of this type of protection is not yet proved and in situations where failure of a device could lead to enormous costs of replacement in complex systems such as computers and radar systems, the risk is often avoided. However, the entire consumer durable market is open to integrated circuit devices, and failure here is not catastrophic and the economics of using a resin-encapsulated device are very attractive. In fact, it has been reported[2] that a U.S. manufacturer has constructed a special plant for the manufacture of encapsulated integrated circuits especially for the consumer goods industry, such as colour television, radio, 'hi-fi', air conditioning and heating.

Selection of polymer for encapsulating components

Epoxy resins dominate the component encapsulation field although the other thermoset resins do offer certain attractive properties. Thermoplastic polymers can also be used and give lower material costs per unit volume and higher production rates than the thermosets. However, the moulding pressures required are normally extremely high (several thousand pounds per square inch) and temperatures of 175 °C and above are necessary. The performance of thermoplastic encapsulants is also poor, cracking and creep occurring in the long term, and maximum operating temperatures are usually limited to about 110 °C. Moisture resistance through the polymer is very high but extremely poor adhesion occurs between the polymer and the lead wires to the device, thus leaving a clear moisture path.

Polyesters were the first casting and potting resins to be used widely for encapsulation. They exhibit good mechanical and electrical properties but suffer from high shrinkage during cure, which can lead to cracking and apparent poor adhesion. They are, however, relatively cheap. Silicone resins, on the other hand, are expensive, but they have superior heat resistance to the other resins and can be used for continuous service at temperatures up to 240 °C. Electrical properties are good, as is moisture resistance, and silicone materials are usually compatible with semiconductor devices because of the absence of impurities in the resin system. With certain grades of silicone, acceptable adhesion to various substrates requires the prior application of a primer to the surfaces concerned, and a further limiting factor with these resins is the evolution of volatile substances which may increase contact resistance in semiconductor devices.

Polyurethanes, like epoxy resins, are extremely versatile and can be formulated to produce an elastomer or a rigid high-strength material. They are non-corrosive and out-gas very little, although the bond that certain types of polyurethane make with substrates can be sensitive to moisture.

Other resinous materials used for encapsulation may be composite systems of two different resins, such as nitrile or vinyl phenolics, nylon epoxy resins or a flexible polysulphide rubber covered, for example, with a rigid epoxy system. Many of the modified phenolics evolve gas on curing, and the nylon epoxy resins are effective only below 100 °C and are sensitive to moisture.

Clearly, it is not possible, nor is it the purpose of this book, to review all the materials used for encapsulation. Nevertheless, it can be seen from the brief comments above that there are very many materials being used to provide protection to electronic equipment, but it is clear that at this stage of development (1974) epoxy resins are the most widely used and versatile of the resinous materials employed. Cured epoxy resins exhibit the following typical properties of importance in encapsulation:

(1) Low cure shrinkage.
(2) Low out-gassing and no volatile substances evolved during cure.
(3) Excellent adhesion to many substrates.
(4) Low water absorption.
(5) Good mechanical and electrical properties.

Moreover, not only is it possible to formulate a vast range of epoxy compounds with widely differing physical, mechanical, electrical and chemical properties, but also to use these systems in any of the multitude of applicational methods employed to encapsulate devices, some of which are listed below:

Liquid resin casting or machine dispensing
Transfer moulding
Liquid transfer moulding
Pre-form moulding
Powder coating
Sealing and dipping.

A selection of these techniques is considered in detail below, illustrating how epoxy resins are used in each of the methods.

Simpson and Pettican[1] suggested that the following principal factors should be taken into account when selecting the most suitable material and process for a particular component encapsulation:

Volume of production.
Permissible capital expenditure on plant.
Acceptable cost of encapsulation.
Degree of moisture protection required (this often determines the thickness of the encapsulant).
Mechanical protection required.
Service temperature range.
Electrical requirements in encapsulant.
Possible effect of encapsulant on component, e.g. chemical interaction, mechanical stress caused during cure, heat dissipation.

Encapsulating techniques

Liquid resin casting or machine dispensing

Small or large numbers of components can be encapsulated equally well either by hand or by using simple automatic metering, mixing and dispensing equipment[7], which can dispense the resin mixture in single shots from 0.5 g to 5 kg or in a continuous stream at up to 5 kg/min. In this technique, it is essential that the mixing operation in the machine should be very thorough so as to ensure proper dispersion of the filler. Vacuum is also applied in order to remove air and is essential if the optimum properties of the system are to be achieved.

The design of formulations for casting and encapsulating uses must take into account many of those factors which are also important in the production of very large castings such as viscosity, absence of volatile substances or contaminants, exotherm and shrinkage. Liquid resin casting is mostly used when high reliability is essential, such as in military or aviation applications, and when costs are of secondary importance. The technique also has some advantages over the most popular encapsulating method, transfer moulding, such as better wetting of the substrates, lower shrinkage and less stress set up during cure[2]. It has certainly been in use much longer than transfer moulding and will continue to be used in many situations. However, when large numbers of miniaturised components need to be encapsulated on a mass production scale, the most popular method has become low-pressure transfer moulding with epoxy resins.

Transfer moulding

This technique consists basically in liquefying a formulated dry epoxy resin powder under heat and pressure and forcing this liquid to flow round the device to be protected, where it is subsequently cured. However, the actual process of transfer moulding is highly complex. The electronic devices that require protection have become smaller and smaller over the years and also more delicate. An integrated circuit may be only 4.5 mm square and have perhaps ten or more leads of diameter 0.025 mm attached to it, and be as complex as a radio receiver. It has therefore been necessary to devise special processes in order to avoid damaging the devices. For example, very low transfer pressures are used, well below the levels used in conventional machines. The surface finish of tools, especially land areas, must also be better than in normal transfer moulding.

Naturally, certain types of device present more difficulties than others during encapsulation, most problems probably occurring with units such as integrated circuits, thin film resistors and glass diodes. Very close control over the temperature, transfer pressure and material flow properties is therefore required in these instances.

Transfer moulding with epoxy resin compounds has enabled substantial cost reductions to be made where massive numbers of components are involved. Butler[8] mentioned that diodes, pulse transformers, pressure transducers, resistors and many other components have been known to fall in price from over £1 each to 20–30 p in 12 months, as a result of mass production encapsulation techniques. Latham[9] pointed out that in the manufacture of diodes and transistors, when fully automatic production lines were established, the cost of the silicon chip was less than the cost of assembly and encapsulation. On the other hand, liquid resin casting is selected when small volumes of components cannot justify the tooling costs involved in transfer moulding, when a high degree of impregnation of the device is required, and when highly tailored formulations are called for. An example of the cost savings possible by using transfer moulding was given for the case of coil encapsulation[10]. By liquid casting, a group of eight girls could cast 2000 coils per shift, whereas the same group was able to produce 10 000 coils in the same time, completely wound and encapsulated, when transfer moulding was employed.

The popularity of the transfer moulding technique in protecting the whole range of devices such as semiconductors, coils, capacitors, resistors, integrated circuits, etc., by encapsulation is due to a number of factors, the most important being:

(1) The moulding compound is an easily handled granular solid, which can be purchased by the moulder ready formulated. There is therefore no need for the moulder to handle resins, hardeners and fillers and compound them together himself.
(2) Standard moulding techniques are very similar to the methods used, and the moulding process is fully automatic. Close control can therefore be maintained over the process, and the number of rejects minimised.
(3) High production rates are possible with low labour costs. Cycle times can be about 15 lifts per hour and cavity numbers can be very high; 120 is common on a 50 ton press.

Epoxy resin moulding compounds also offer great latitude in formulation and therefore a wide range of characteristics[10], and require very low moulding pressures.

Clearly, for the successful exploitation of this technique, considerable expertise is also required for the design of transfer moulding machines and moulds for them, but these aspects fall outside the scope of this book. Further details on these matters can be obtained from the very large number of papers published in recent years (for example, see refs. 11–16).

Resins other than the epoxy resins, such as silicones, alkyds and phenolics, can also be used in this technique. In order to understand why the epoxy resins have achieved their pre-eminent position in this market, it will be useful to consider the properties desirable in transfer moulding compounds and some typical actual values. *Table 10.1* lists some of these data[17]. Considering first the moulding properties of the various compounds, the important characteristics are concerned with pressure, temperature and time. A long and soft flow is essential, together with a low transfer pressure, such as a few hundred pounds per square inch. Pressures higher than 1000 lbf/in^2 may remove insulation from wires as well as result in poor impregnation and distortion. The units being encapsulated may be displaced, which could result in non-uniform wall thicknesses, which might lead to eventual breakdown of the component in service. Moulding temperatures are usually about 150–160 °C and cannot easily be increased above 200 °C because of the effect on the other materials in the device. The ability to encapsulate at 200 °C or below is an advantage over ceramic or glass encapsulation, which require a temperature of 800 °C. The silicon chip can only withstand a temperature of 400 °C[9]. The moulding time should also be as short as possible in order to minimise costs per unit. Frequently, the resin is not fully cured when the units are ejected from the press, and final curing is carried out separately in an oven.

The formulation of a transfer moulding powder must strike a balance between all those factors which can affect the flow and gel times of the powder. Values of these properties which are high or low can have adverse effects on the devices being protected. Although epoxy resins are the most widely used resins for transfer moulding, other polymers can be and are used, although each suffers from certain drawbacks when compared with the epoxy powders. For example, phenolics are relatively cheap resins and flow well under high pressures. At the much lower pressures used in transfer moulding they can out-gas ammonia, water and also free phenol, liberated in the resin during the curing reaction[10]. In addition, the adhesion of cured phenolic resin systems to inserts is poor. The drawbacks to the use of the silicone resins are different from those of the phenolics. The silicones are expensive, require fairly long cure times (4–5 min) and the final mouldings have poor mechanical properties. They are, however, often used because of their good flow at low moulding pressures, perhaps 300 lbf/in^2, and the excellent high-temperature properties of the mouldings. The stability of the silicones is good, and they are usually compatible with semiconductor devices owing to the purity of the resins.

Table 10.1

PROPERTIES AND DESIRED VALUES FOR PLASTIC ENCAPSULANTS (AFTER DERRINGER[17])

Property	*Desired value*	*Glass-filled encapsulating epoxy*	*Glass-filled DAP*	*Glass-filled silicone*	*Glass-filled phenolic*	*Mineral-filled alkyd*	*Nylon 6*	*Poly-sulphone*
Moulding:								
Pressure, p.s.i.	1000 or lower	50–1000	500–4000**	1000–5000**	1000–6000**	100–4000**	10 000–25 000*	1000
Temperature, °C	149 °C	138–165**	132–165**	154–182*	138–177**	138–160**	227–316*	288–316*
Time	As short as possible							
Physical:								
Tensile strength, p.s.i.	8000 minimum	5000–15 000**	8000–11 000	4000–5000*	5000–18 000**	3000–8000**	7000–12 000**	10 200
Izod impact, ft lb	0.3 minimum	0.5	0.5	3–15	0.3	0.3	1.0–5.5	0.3
Heat resistance, °C	+150	150–230	177–230	316	177–288	150–230	80–120*	120–174**
Flame resistance	Self-extinguishing	Self-extinguishing to non-burning	Self-extinguishing to non-burning	Non- to-slow-burning	Non-burning	Non-burning	Self-extinguishing	Self-extinguishing
Thermal conductivity 10^{-4} cal/cm/s/deg C	High as possible	4.10	5.15	7.51–7.54	9–14.5	15–25	5.85	6.0
Electrical:								
Dielectric constant	4.75 maximum	3.5–5.0**	4.1–4.4	3.2–5.0**	6.9*	5.0–6.2*	4.0–4.9**	2.58–3.13
Dielectric strength, V/mil (short time)	300 minimum	250–400**	395–450	200–400**	140–400**	350–450	440–510	425
Dissipation factor, 1 kHz	0.015 maximum	0.01	0.004–0.009	0.003–0.025**	0.02*	0.007–0.03**	0.01–0.06**	0.0011

* Indicates values outside of desired specification.

** Indicates question areas even though broad family properties may lie within specification. (Selection of a moulding compound within the family must take this into account.)

Diallyl phthalate transfer moulding powders, which can be classified with the unsaturated polyesters or alkyd resins, are also used but moulding pressures of several hundred pounds per square inch are required in order to obtain good flow. The dimensional stability of DAP mouldings is good, but their adhesion to inserts is poor and tensile and impact strengths are moderate.

Thermoplastic materials, such as the polysulphones and nylon 6, can also be formulated into transfer moulding powders that are considerably cheaper than the thermosetting powders. However, they are usually processed at unacceptably high pressures, show cracking and creep after ageing at elevated temperatures and show a virtual absence of adhesion to lead wires and other surfaces, hence nullifying their outstanding water resistance and price advantage.

The formulation details for epoxy transfer moulding powders are normally not released by their formulators, who have usually invested a great deal of time and money into their development, but the principles underlying the formulation of epoxy moulding compounds (both transfer and compression) have been well enumerated by Flynn and Cianciarulo[18] and typical properties of some epoxy transfer moulding systems were given by Byrd and Saltzman[19].

The questions of thermal expansion and shrinkage mentioned above are important when considering possible failure modes for encapsulated devices. Synthetic resins have much greater coefficients of thermal expansion than metals and, although fillers can reduce the coefficient to some extent, they do not reduce it to the level of coefficients for metals. Moreover, at temperatures of 100–150 °C, the cured resins may begin to soften, and at 200–250 °C they may decompose, depending, of course, on the system considered. This movement of the protective medium may lead to a loss of adhesion with the leads, and hence create a path by which moisture can enter the component[9]. Excessive shrinkage during cure can lead to the same situation. Permeation of water directly through the resin itself, although possible, is unlikely to be a prime cause of failure.

Finally, looking to the future, there is no doubt that more manufacturers will turn towards transfer moulding as a technique for protecting electrical and electronic devices with synthetic polymers. Epoxy resins will also continue to be challenged for their leading place by newer thermoplastics such as polyacetals, polycarbonates, polysulphones and polyamides, and the thermosetting polyurethanes will also be a strong contender in this market. But the tremendous choice of formulation possibilities available with epoxy resins will always ensure that they obtain a major proportion of this market.

Liquid transfer moulding

This new technique seeks to combine some of the advantages of liquid casting with the high production rates obtainable with transfer moulding. It consists of injecting under pressure a liquid epoxy resin formulation into a multicavity mould that is held under pressure by a clamping device such as a hydraulic press. The resin system can either be stored ready mixed in a tank, if the pot life permits, or metered, mixed and dispensed automatically into the machine at the beginning of the moulding cycle. The usual range of thermosetting resins can be used, such as the silicones, polyurethanes and polyesters, but most work has been carried out with liquid epoxy resin systems.

Bell and Zecher[20] and Uglione[21] have given full descriptions of the process and claim for it the following advantages over other encapsulation methods:

(1) Liquid transfer moulding can be carried out at temperatures as low as 90 °C and pressures down to 5 lbf/in^2, although 35–50 lbf/in^2 is more usual. This allows clamping pressures to be lower than with normal transfer moulding, hence simplifying mould design and greatly reducing the cost of equipment.
(2) Conversely, a greater area of moulding can be made if the same pressure as in a transfer moulding press is used.
(3) A further advantage of the ability to mould at low pressure lies in the possibility of very delicate components, such as thin and thick films, glass substrates, integrated circuits and complex high density circuits, being encapsulated.

At present, a number of component encapsulations are being considered for this method[21], including integrated circuits, automotive coils, transformers, resistors, capacitors and light-emitting diodes. It is claimed that certain coils could be encapsulated only by liquid casting or liquid transfer moulding techniques in order to avoid distorting the windings or the bobbins. If the coils are to be mass produced, liquid transfer moulding is to be preferred as a cycle time of less than 3 min at 300 °F (150 °C) is possible. Similar considerations have led to the choice of this technique in the other examples quoted.

Pre-form moulding

In this method, pre-formed pellets of uncured or partially cured (B-staged) resin systems of various shapes and sizes are used often in conjunction with moulded plastic cases or sleeves. On heating the pellets, the mixture melts and flows into the mould or casing of the component, where it gels and is then cured to its final state. No mixing or handling of the resins, curing agents and fillers is necessary, no moulds are required to be cleaned and no resin is wasted. In addition, only a small amount of inexpensive equipment, such as ovens and jigs, is required. The finished product is of good appearance and uniform in shape and size.

The method is used for encapsulation, sealing, end-sealing and embedding in metal cans, and a range of pellets is available in different shapes and sizes.

Powder coating

The use of this technique for the insulation of small electric motors and other components has already been described (Chapter 7). The use of epoxy powders in fluidised beds to coat pre-heated electronic components is also taking place, and can be expected to increase in the future. The method is, of course, limited to components that can withstand the temperature at which the powders are cured (120–200 °C).

PRINTED CIRCUITS

In most branches of the electronics industry, printed circuits have now replaced wiring. This change, in which the electric current is carried not in wires but in a metallic foil which is integral with the insulating board, has revolutionised the industry and led to the manufacture of printed circuit boards becoming a separate industry in its own right.

The chief virtues of the printed circuit board have been described[22] as:

(1) It satisfies the structural and electrical requirements of the assembly of which it forms a part.

(2) It enables savings in weight and space to be made.

(3) It facilitates the design of equipment.
(4) It helps to reduce costs by enabling standardisation and automated production to be adopted, and these techniques require fewer skilled staff.
(5) Inspection and fault correction are easier.
(6) It facilitates mass production and miniaturisation.

Most circuit boards for use in domestic television and radio receivers and for non-critical industrial applications are made from paper- or cloth-reinforced phenolic resins. These boards are easy and cheap to make, with a low reject rate in production, and are completely adequate for these less demanding situations. Glass–epoxy laminates are used in industrial applications such as computers, telecommunication systems, television equipment including colour receivers and missile and radar guidance electronics and navigational equipment, where high performance and reliability are required from the circuit board. The largest user of epoxy boards is probably the computer industry; it has been estimated[22] that this industry takes about 75 per cent of the board produced. Epoxy boards are also being used in increasing amounts in multi-layer circuit arrangements. In Europe, epoxy–glass board probably held 40–45 per cent of the total market for circuit boards in 1972 and will probably continue to increase its share, perhaps to 75 per cent of the total by the late 1970s. Naturally, printed circuit

Table 10.2

PROPERTIES OF PRINTED CIRCUIT LAMINATES BASED ON FOUR DIFFERENT RESINS (AFTER DRAPER[23])

Property	*Units*	*Phenolic*	*Melamine*	*Silicone*	*Epoxy*
Tensile strength	p.s.i.	25 000	40 000	25 000	40 000
(BS 2782/301C)	MN/m^2	175	280	175	280
Flexural strength	p.s.i.	30 000	55 000	28 000	65 000
(BS 2782/304B)	MN/m^2	210	385	196	455
Water absorption (BS 2782/502D)	mg (0.062 in thick)	20	30	2:0	5.0
Insulating resistance (BS 1137 App.H)	Mohm	10^4	10^3	10^5	10^5
Tan δ at 10^6 Hz		.0.025	0.020	0.002	0.020
Permittivity at 10^6 Hz		5.5	7.8	3.9	5.0
Relative costs:					
paper-based		1.0	–	1.2	1.5
glass-based		–	–	6.0	4.0

boards can be made from laminates based on thermosets other than epoxy resins and phenolics, for example silicones and melamine resins. Compared with paper–phenolic boards, epoxy–glass boards are in general stronger, more resistant to hot solder and have better warp resistance and electrical (anti-tracking) properties. Taking a broader view and considering the whole spectrum of resin boards, epoxy laminates are no doubt the best choice from the point of view of mechanical, thermal and electrical properties. Even when cost is considered, epoxy laminates are the best choice when high performance is needed. *Table 10.2* illustrates these statements by referring to the different performances that can be obtained from glass–cloth laminates made with four different resins[23].

The important properties of the epoxy–glass laminates can be summarised as follows:

Physical and mechanical properties

(1) High flexural and tensile strength.
(2) Low moisture absorption.
(3) Good dimensional stability.
(4) Resistance to heat ageing at moderate temperatures.
(5) Fair machineability.
(6) Good bond to copper circuit foil.
(7) Fire retardant (brominated grades).
(8) Good colour.

Chemical properties

(1) High chemical resistance to fluxes, solvents, etchants, plating solutions and molten solders.

Electrical properties

(1) Low dielectric constant and dissipation factor values at low and high frequencies.
(2) High dielectric strength.
(3) Good insulation resistance.
(4) Fair arc and tracking resistance.

In the U.S.A., the National Electrical Manufacturers Association (NEMA) has published specifications for a variety of laminated products including printed circuit board, and the NEMA specifications for copper-clad epoxy laminates are virtually identical with the U.S.

Military (MIL-P) specifications for these materials. The NEMA specifications are well known and accepted as primary standards throughout the world and cover the following grades as far as epoxy resins are concerned:

G-10	general purpose.
FR-4	fire retardant.
G-11	general purpose elevated temperature service.
FR-5	fire retardant elevated temperature service.

G-10 and FR-4 grades

Many epoxy resin systems could be used to prepare laminates that meet this general purpose specification, and in 1970 about 50 per cent of all epoxy–glass boards made were of G-10 grade. Similarly, about 30 per cent were of FR-4 grade, and the trend is towards increasing use of this grade at the expense of G-10. Both boards have good punching and shearing properties and machining characteristics and a reasonable level of other properties at room temperature. The main disadvantages of the G-10 grade are poor resistance to molten solder and a decrease in properties at elevated temperatures. The FR-4 grade, compared with G-10, shows a lower adhesion to copper, lower insulation resistance and higher raw material costs.

Originally, mainly for technical and economic reasons, the dicyandiamide (DICY) cured low molecular weight solid DPP–ECH resin (for example, Epikote 1001 from Shell) was generally favoured to meet the G-10 specification. Now, major resin manufacturers have special grades tailor-made for the 'pre-preg' laminating process. All of the epoxy laminated boards which are made for printed circuit use are made by the pre-impregnation method. Improvements in the elevated temperature properties of this system can be made by replacing about 20 per cent of the resin with an epoxy novolac resin. *Table 10.3* illustrates the improvement in flexural strength retention at 100 °C that this substitution can achieve.

Laminates meeting the FR-4 specification can also be prepared using DICY as a curing agent in conjunction with a brominated bisphenol-A epoxy resin. Improvement in the elevated strength retention can also be made by replacing about 20 per cent of the halogenated resin with an epoxy novolac resin, without significantly reducing the flame retardant characteristics.

Table 10.3

FLEXURAL STRENGTHS OF LAMINATES (AFTER SHELL CHEMICAL COMPANY[24])

Laminating system	*Resin content,* per cent	*Flexural strength of laminate,* lbf/in² (kg/cm²)		
		At room temperature	*At 100 °C*	*Retention of flexural strength,* per cent
Epikote 1001–DICY	36	86 600 (6000)	16 000 (1100)	19
Epikote 1001–epoxy novolac–DICY	41	60 000 (4200)	40 000 (2800)	67

G-11 and FR-5 grades

These laminates are used when class B (130 °C) insulation performance is required, and in 1970 they accounted for about 10–20 per cent of world-wide epoxy circuit board production. The mechanical and electrical properties are similar to those of the G-10 and FR-4 grades at room temperature with the additional advantage that 50 per cent of the initial flexural strength is retained at 150 °C.

The disadvantages of these grades compared with the G-10 laminates are poor machining characteristics, lower adhesion to copper and higher raw material costs.

Future developments

While laminate manufacturers consider it unlikely that sought for improvements will be sufficiently important to justify an increase in the price of the resin system employed, progıess made in meeting these improvements would give a resin manufacturer a keen competitive edge over his competitors. These technical changes being sought are as follows.

(a) Higher HDT

An increase in the ability of laminates to withstand higher temperatures is desirable. A minimum HDT of 155 °C, without any increase in brittleness or fall-off in copper peel strength, would improve resistance to thermal shock, give better resistance to blistering on immersion in heated oils and increase resistance to resin 'smearing' on high-speed drilling. This latter condition is the softening of the resin on the wall of a drilled hole, which causes difficulty in depositing copper for through-hole plating.

(b) Flame retardance

The use of brominated resins is the widely accepted method of obtaining flame retardant laminates. However, the hydrogen bromide evolved from the resin in a fire is highly corrosive. Furthermore, the presence of bromine in the resins can reduce copper peel strength and lower long-term breakdown and dielectric strength under humid conditions. Any alternative flame retardant system without these disadvantages would therefore be well received.

(c) Humidity resistance

An improvement in this property is sought, with the hope that this would stabilise electrical performance.

(d) Dielectric constant and dissipation factor

The NEMA G-10–G-11 specifications require board to have a maximum dielectric constant of 5.4 at 1 MHz and a dissipation factor of 0.035. Although most commercial boards have dielectric constants of 4.8–5.2, a value of 4.6 and a dissipation factor of 0.01 are required.

(e) Strength retention at elevated temperatures

The G-11 and FR-5 laminates will be required to retain 75 per cent of their initial flexural strength at 150 °C, rather than the present 50 per cent.

(f) Punchability

It would be desirable to punch holes in laminates in preference to drilling. However, holes of diameter less than 0.032 in cannot be punched in existing laminates without surface cracking. Nevertheless, easy punchability is required although it would be more difficult to achieve if the HDTs increase.

(g) Copper peel strength

Some laminators would like to see the copper peel strength of the board increased from 8–12 to 15–20 lbf/in^2 (from 56–84 to 105–140 kN/m^2).

REFERENCES

1. SIMPSON, W.E. and PETTICAN, K. A., *Electronic Engng,* **42**, No. 503, 50 (1970)
2. MANFIELD, H. G., *BEAMA Elect. Insulation Conf. Proc.* (1970), Paper 7a
3. MUELLER, B. H. and HARPER, C. A., *Elect. Mnfr,* **65**, No. 2, 119 (1960)
4. HONSINGER, V. B., *Elect. Mnfr,* **64**, No. 5, 114 (1959)
5. CRAWFORD, D. E., *Mater. Design Engng,* **46**, No. 1, 99 (1957)
6. JEWELL, J. M. and JENNER, W. C., *Mod. Plast.,* **37**, No. 4, 101 (1959)
7. Shell Chemicals Technical Literature
8. BUTLER, M. J., *Br. Plast.,* **43**, No. 10, 135 (1970)
9. LATHAM, G. R., *Electronic Components,* **8**, No. 5, 518 (1967)
10. *Mod. Plast.,* **43**, No. 2, 96 (1965)
11. BINDER, H., *Insulation, Lake Forest,* **14**, No. 3, 54 (1968)
12. WALKER, K. G., *Electronic Components,* **9**, No. 9, 1009 (1968)
13. ZEIDO, J. E., *Mod. Plast.,* **47**, No. 6, 116 (1970)
14. ZECHER, R. F., *Insulation, Lake Forest,* **13**, No. 1, 37 (1967)
15. PISCHEL, O. H., *Kunststoffe,* **58**, No. 4, 1(1968)
16. *Papers presented at the 28th ANTEC meeting of the SPE, U.S.A.* (May 1970)
17. DERRINGER, C. K., *Insulation, Lake Forest,* **16**, No. 2, 38 (1970)
18. FLYNN, R. and CIANCIARULO, A. N., *SPE Jl,* **24**, No. 11, 37 (1968)
19. BYRD, P. S. and SALTZMAN, R., *Mod. Plast.,* **46**, No. 11, 128 (1969)
20. BELL, F. and ZECHER, R. F., *SPE Jl,* **25**, No. 2, 40 (1969)
21. UGLIONE, H. L., *Polym. Engng Sci.,* **10**, No. 6, 345 (1970)
22. *Araldite in Printed Circuits, Part 1,* CIBA-Geigy
23. DRAPER, C. R. (Editor). *Printed Circuits and Electronic Assemblies,* Robert Draper, Teddington, Middlesex (1969)
24. Shell Chemicals Technical Literature

SECTION IV
Construction Industry

Introduction

Over the centuries, the underlying principles of the methods used in construction have not changed greatly. The techniques used in building are common to most societies around the world, although they have developed more rapidly in some countries than in others. Amongst the many factors which determine the techniques that are used and their rate of development, the local climate and the local materials most readily available have played important parts. The primary building materials are concrete, wood, brick, asphalt, iron and steel, and are, in general:

(1) Cheap and plentiful.
(2) Usually adequate in performance for the job.
(3) Found widely throughout the world.
(4) Familiar to local craftsmen and can be manipulated and worked by simple well known methods.

These are good reasons for the continuing use of these materials and any new material has to provide clear benefits to the industry before it will be accepted.

The nature and structure of the construction industry differ from most other industries. The work is mostly carried out in the open, the 'workpiece' being immovable, and as workmen come and go the product of their labour remains fixed at one point. In most countries, it is usual to find large groups operating on the civil engineering side of the industry, whereas on the domestic and light building side there are still very many small firms. These latter firms are frequently under-capitalised, resistant to the introduction of new ideas and served by a wide variety of manufacturers and organisations. Architects and designers have to decide upon the choice of new materials and methods from a welter of claims from different suppliers. In this situation, it is to be expected that it will take a long time for new materials such as plastics to become accepted and for the workforce to be properly trained in their use. The need to have confidence in the long-term performance of new materials is also an important factor which inhibits their rapid acceptance. The history of the use of epoxy resins in the construction field is no exception to these generalisations. However, as costs, especially labour costs, rise steadily in all countries and become an ever-increasing proportion of building costs, the need to develop new methods and materials that can effect cost savings becomes more pressing.

Initially, epoxy resin compositions were only used remedially, and this remains a major use. Increasing amounts of resin are now being used in new work although so far the resins have not replaced traditional materials to any significant extent in primary construction use. Rather, they have been used to improve the performance of traditional materials or to obtain special effects and benefits not otherwise achievable, or to allow new techniques to be used or old ones improved. No doubt in the future this trend will continue and epoxy resins will find their way in even greater amounts into primary structures. Certainly, the use of epoxy resins in the construction industry in its widest sense is growing faster than any other outlet for the resins. The following chapters in this section review the major applications that are accounting for this growth.

11 Flooring

Epoxy resin flooring is used as the covering or topping that is frequently placed on top of a sub-floor. This covering, which is, of course, the wearing surface, can be classified in a number of ways. The composition of the floor, its method of laying and the nature of the wear to which the floor will be subjected have all been used as ways of grouping different floor types. It is sufficient when considering epoxy resin floors merely to divide the floors into three types:

(1) Domestic floors, as in dwelling houses and flats. Examples of such floor materials are carpet, PVC, wood, linoleum and cork, each in tile or strip form.
(2) Institutional floors in schools, offices, banks, hospitals, showrooms and certain types of factories. Many of the materials used are the same as those for domestic floors, such as PVC, wood and linoleum, and in addition other materials, such as concrete, magnesite, ceramic tiles and terrazzo are employed.
(3) Industrial floors, as in certain types of factory where severe corrosion and/or heavy mechanical wear occurs, e.g. sweet-stuff and food factories, tanneries, breweries, battery and plating shops and chemical plants. Examples of the flooring used include acid-resistant quarry tiles, metal paving, granolithic materials, asphalt and synthetic polymers.

All of these types of floors have similar basic requirements, but the relative importance of each requirement is different for each type of floor. This may be shown in the form of ratings as in *Table 11.1*. Naturally, there will be exceptional situations where the assessment given in *Table 11.1* is not appropriate for that particular example, but in general terms it is believed that this overall assessment is essentially accurate.

The data show that decorative appearance and good acoustic insulation properties are of prime importance for domestic floors. For institutional floors, non-slip characteristics and the ability to clean the floor thoroughly are the most important properties, followed by good sound insulation. This last factor is becoming increasingly important as the number of industrialised buildings increases, because sound transmission is usually high in this type of building. In general, there is increasing awareness of the nuisance and distraction caused by high noise levels, which result in reduced efficiency. The most direct way of reducing noise is to use a sound-deadening covering on the floor and the trend has therefore been towards the use of soft flooring. Industrial floors, as would be expected, should primarily be resistant to mechanical and chemical attack, be capable of withstanding thorough, vigorous cleaning, and have non-slip characteristics.

Embracing all of the factors considered in *Table 11.1* is cost. In most instances, the material which satisfies the performance requirements at the lowest cost will be used. In general, flooring material that has a lower performance rating than average in a given area coupled with a high cost can be disregarded.

Table 11.1

RATINGS OF FLOOR REQUIREMENTS

Property	*Order of preference (1 = most preferred)*		
	Domestic	*Institutional*	*Industrial*
Decorative: Pleasing appearance Pleasing feel Choice of colour and pattern	1	4	4
Good sound insulation	2	2	5
Easily cleaned	3	3	3
Functional safety: Non-slip Capable of washing to sterile condition (crack-free)	4	1	2
Functional service: Resistant to trucking Resistant to chemicals	5	5	1

DOMESTIC FLOORING

Considering the five groups of properties in *Table 11.1*, one study which compared typical domestic floor finishes with an epoxy finish gave the following ratings (the lowest figure represents the flooring with best performance properties):

Carpet	5
Vinyl (foam backed)	9
Wood parquet	15
Linoleum	15
Thermoplastic tiles	17
Epoxy	20

Epoxy flooring on this basis is the least desirable for domestic purposes, chiefly because of its low level of sound insulation and its lack of pleasing appearance. Efforts have been made to improve epoxy resins in these respects, for example by using mixed coloured aggregates or spattering colour on to the floor surface to improve the appearance, and by incorporating rubber crumb as an extender to improve the flexibility and 'feel' of the floor and to help to reduce impact noise. However, these modifications have not been successful and really give a poor imitation of vinyl flooring at higher cost.

In general, carpet and wood parquet are more expensive than epoxy flooring, but they clearly have much more appeal in the domestic market. For epoxy resins to enter the domestic market, therefore, they would need to compete with the cheaper vinyl and linoleum flooring, whose performance characteristics are nearer to that of epoxy flooring than any other domestic floor. Unfortunately, these materials have such a low price compared with epoxy resins that even the development of new, faster ways of laying epoxy flooring will not overcome the difference.

Clearly, on both technical performance and cost, epoxy floors are not likely to succeed in the domestic market, which is moving strongly towards the use of soft flooring and carpets.

INSTITUTIONAL FLOORING

Institutional floors include a wide range of flooring situations, from 'light duty' floors for offices, lecture rooms and corridors to floors for industrial situations such as laboratories, kitchens in hospitals, schools, hotels, showers and bathrooms, etc. It is therefore more convenient to consider the latter group of floors under the heading 'industrial flooring' in the next section.

The institutional floors that remain from the above list do not offer a very promising market for epoxy resins, despite earlier optimism that the self-levelling epoxy or epoxy–terrazzo would be used widely in office foyers, internal concourses of large office buildings, etc. The order of preference for the different types of floors is probably as follows:

Carpet	7
Vinyl	12
Wood parquet } Epoxy }	14
Linoleum	15
Thermoplastic tiles	16

At present, low-cost linoleum and vinyl and thermoplastic tiles account for about 70 per cent of the total area treated, but on a performance basis, carpet is clearly the preferred flooring material. Compared with the harder flooring, carpet has improved comfort, sound insulation and, it is claimed, a 10 per cent advantage in cleaning costs. It therefore seems unlikely that epoxy flooring will be preferred to carpet or vinyl tiles.

INDUSTRIAL FLOORING

Most industrial sub-floors are made from concrete, steel or wood and are either left uncovered or are covered with a variety of materials for comfort, functional or aesthetic reasons. Concrete is probably the most widely used industrial floor material and is frequently left uncovered. However, concrete 'dusts' throughout its life and readily absorbs fats and oils that are spilt on it; it abrades easily and is not very resistant to corrosive materials. It is therefore often necessary to cover concrete and other industrial floors with a topping in order:

(a) to protect the sub-floor from corrosive materials;
(b) to protect the sub-floor from severe mechanical wear;
(c) to allow high standards of cleanliness and freedom from dust to be obtained;
(d) to provide a non-slip surface, which preferably has good sound insulation and a pleasing appearance.

The choice of an industrial floor covering to provide these properties involves consideration of factors such as the physical and mechanical properties, chemical resistance, ease and cost of laying and of maintenance of the covering[1].

A very wide choice of materials for industrial floors is available in two basic types:

(1) Prefabricated tile blocks or slabs laid in a wet cement or sand screed with a chemically resistant mortar grout or a bitumen adhesive.
(2) *In situ* monolithic floors.

In type (1), the tiles can be fully glazed ceramics, quarries or metal paving and the blocks wooden when a light industrial floor is required,

Table 11.2

RATINGS OF THE PROPERTIES OF DIFFERENT INDUSTRIAL FLOOR TYPES

Property	*Order of preference (1 = most preferred)*							
	Concrete	*Bitumen cement*	*Asphalt mastic*	*Granolite concrete*	*Resin-modified concrete*	*Ceramic tiles*	*Wood block*	*Epoxy*
Functional service:								
Resistant to trucking	6	7	8	4	3	1	5	2
Resistant to chemicals	6	4	1	5	2	1	3	2
Functional safety:								
Non-slip	4	4	3	4	4	4	2	1
Suitability for sterile cleaning	3	3	3	3	2	1	4	2
Easily cleaned	3	3	3	2	2	1	4	1
Decorative:								
Pleasing appearance	4	5	5	4	4	1	2	3
Pleasing feel	3	3	2	3	3	3	1	3
Choice of pattern and colour	4	5	5	4	4	1	2	3
Good sound insulation	3	3	2	3	3	3	1	3
Cost, £/m² laid 1970	0.75	1.00	1.00	1.78	2.50	3.00–7.00	5.00	1.78–7.00

whereas type (2) floors can be composed of granolithic cement, asphalt or synthetic polymers. Epoxy floors are mostly of type (2), i.e. the continuous seamless floor, although there has been a limited development of tiles made from epoxy resins. Epoxy resins can also be used as tile grouts (Chapter 13).

An analysis of the order of preference for each property for the various floors, as in previous sections, is given in *Table 11.2.*

Summarising the ratings of the first five (most important) properties in *Table 11.2* leads to the following order of preference:

Floor	Rating
Ceramic Epoxy	8
Resin-modified concrete	13
Wood block Asphalt mastic Granolithic concrete	18
Bitumen cement	21
Concrete	22

Epoxy floors are clearly competitive on a performance basis with all the floor types considered. On a cost/performance assessment, epoxy floorings remain competitive with ceramics, but are probably not competitive with the cheaper finishes such as asphalt mastic and granolithic concrete, where the performance of these systems is acceptable. Moreover, despite the trend towards the use of higher quality floors in factories, the floor is one of the first items that is cheapened when a lower overall cost has to be achieved by a contractor. Nor is there likely to be a serious trend towards replacing cheap by more expensive floors.

It therefore appears that the main market for epoxy floors will be the high-quality industrial floor market that has been traditionally occupied by ceramic tiles. The advantages that epoxy floors can offer in this application include improved non-slip characteristics and cleaning ability, and these are discussed more fully below.

The final choice of a particular floor covering can be decided only in the light of the foreseen in-service conditions to which the floor will be subjected. Factors such as the nature and frequency of chemical spillage, the type and extent of mechanical wear and the nature and frequency of cleaning, together with the maximum variations in temperature, are the most important variables to be taken into account. Considering the various types of tiles and other discrete forms of flooring, fully vitrified ceramic tiles can be made completely acid-resistant, non-slip, resistant to impact and impervious to chemicals and similarly, quarry tiles, bricks and paviours and metal tiles are all fully capable of withstanding heavy mechanical wear and certain chemicals. However, they are all heavy, expensive to lay and are accompanied by large areas of jointing, which can be vulnerable to chemical attack, even when special grouting cements are used. Tiles are also cold, noisy, often slippery when wet and are not easily laid round machinery, pillars and other obstructions. The bedding and jointing mortars used often take several days to harden and the floor therefore cannot be used during this period.

In situ monolithic floors are jointless, lighter in weight than tiles, quicker to lay and can easily be laid round obstructions. They can be made non-slip and chemically resistant and can withstand heavy mechanical wear. However, they are mostly cold, hard and noisy and sometimes require lengthy preparation of the sub-floor before laying. Granolithic toppings (cement, granite chips and sand) are fairly cheap but are attacked by weak acids and oils and require 7 days to cure. Floors based on thermosetting resins were originally developed with epoxy resins and were jointless floors of two types, screeded and self-levelling. Polyester-based floors were then developed and, more recently, polyurethanes have been used for flooring. The advent of the latter materials also introduced the idea of flake poly(vinyl acetate) (PVA) used in combination with a polyurethane resin to give a decorative floor. All three thermosetting resins can be used as the binder in a floor system that is filled either with aggregate or flake PVA. What, then, are the relative advantages and disadvantages of using epoxy resins rather than the other resins[2]?

All three types require good ventilation when they are being applied. In addition, the floor temperature should not be below about 10 °C so as to ensure that retardation of cure does not occur. This effect is the least critical with polyurethanes. When the temperature exceeds about 25 °C, the cure is accelerated, especially with polyurethanes under humid conditions. All of the resins shrink during cure, polyesters showing the greatest effect (6–7 per cent volume reduction).

Polyurethanes cure from the surface downwards and carbon dioxide must therefore not be trapped, or else bubbling will occur. The rate of cure of moisture-cured polyurethanes is dependent upon the absolute atmospheric humidity, which can be a limiting factor to the early use of a floor after laying. Pigmentation can also be difficult with the polyurethanes, as can stability to ultra-violet light, this latter property also being a matter of concern with amine-cured epoxy resins.

Polyesters have good rheological properties and flow readily, but suffer from cure inhibition and bad adhesion to damp substrates and are also sensitive to alkaline conditions. Both epoxy resins and polyurethanes are superior to polyesters in these respects. Polyurethanes can also combine flexibility with hardness and produce a 'quiet' floor. Epoxy resins are more rigid than the polyurethanes but are very resistant to attack by alkalis, whereas attack by acids is withstood better by polyesters.

In summary, it does not appear that polyesters or polyurethanes offer serious competition to epoxy resins in trowelled floor applications. In fact, there are clear performance advantages with the epoxy systems as far as laying characteristics and the ability to hold aggregate under high wear conditions are concerned. However, in light-duty applications, polyurethane floors might present competition if an alternative to the PVA flake decorative effect is offered.

TYPES OF EPOXY FLOORING

There are three types of epoxy flooring systems:

(1) Pourable self-levelling seamless floors.
(2) Trowelled floors.
(3) Terrazzo floors.

In addition, epoxy adhesive primers may be used with other overlays, e.g. with polyester or polyurethane overlays. These epoxy flooring systems can be used as new floor coverings over a sub-floor of concrete, steel or wood and can also be used remedially and applied over existing floors. The replacement or repair of a damaged floor can be very expensive. The average life of a floor is about 10 years[3], but when exposed to abrasive wear, attack by chemicals, vigorous cleaning methods or the impact of heavy objects, a floor may need to be replaced much more frequently, even as often as every 6–12 months. Epoxy flooring compounds are 'two-can' systems and consist essentially of the resin together with ancillary chemicals, such as diluents or flexibilisers, in one can and the curing agent in the second can. If a pigment paste is to be used to produce a coloured floor, it is added to the resin before mixing with the curing agent. When aggregates (fillers) are to be used, they are mixed with the binder after the resin and curing agent components have themselves been thoroughly mixed together. By suitable formulation, a range of colours, surface finishes and thicknesses can be obtained to meet specific requirements.

Important properties of epoxy resin floors are as follows:

(1) High resistance to a wide range of acids, alkalis, solvents, fuels and other corrosive materials.
(2) They are laid as continuous, jointless screeds or toppings, so that there can be no seepage through joints to the sub-floor or room below. They can be readily formed around machinery and shaped into channels, gullies, skirtings, etc.
(3) They can be laid quickly with the minimum interruption of work and alteration to existing floor levels.
(4) They can cure rapidly (12–24 h) and down-time for machinery is therefore minimal.
(5) The floors are highly resistant to abrasion and virtually eliminate the problem of dust formation. They also have high mechanical strength and favourable strength-to-weight ratios, especially in comparison with concrete.
(6) They adhere with immense strength to all common flooring substrates when the proper surface preparation is carried out.
(7) They can be produced with a non-slip surface that is effective even when wet and they are readily cleaned.

The difference between the three types of epoxy floors, apart from cost, is their suitability for different types of application. Heavy-duty applications generally require trowelled or terrazzo floors, the self-levelling systems not having the necessary level of performance needed for such conditions. However, self-levelling types have higher laying speeds and lower costs, and are often preferred for floors in laboratories, computer rooms, telephone exchanges, etc., where non-slip, chemical resistance and freedom from dusting are required, but heavy wear conditions do not occur. Heavy-duty trowelled floors, in contrast, find application in industrial processing areas where chemical spillage and/or heavy trucking is involved or where, for example, grease contamination can cause serious skid hazards.

Self-levelling floors

These flooring systems consist of a free-flowing mixture with very little filler content, which is applied either by spray or simply by pouring it on to the prepared sub-floor. The composition can then be spread out with a plastic comb or similar implement to an overall thickness of about 1/16 in (1–2 mm). The sub-floor must therefore be sufficiently even to accept this thickness of topping. The binder system can be used unfilled, which results in a floor with an impervious high-gloss finish. Alternatively, a filler, such as silica flour, can be incorporated at a maximum ratio of 1.5 : 1 part by weight of resin binder. The non-slip characteristics of these floors can be improved by scattering on to the wet surface a fine aggregate such as carborundum or calcined bauxite. Alternatively, the hardened floor can be lightly abraded. A decorative effect can be obtained by incorporating PVA paint flakes into the epoxy binder and subsequently sealing the floor with a polyurethane varnish. This type of floor has occasionally to compete with linoleum, PVC tiles and sometimes carpets, and in these situations the chemical resistance properties of epoxy resins are obviously not of great importance. The wear characteristics of the floor are at least as good as those of the highest quality PVC or linoleum and the floor can also withstand severe foot traffic and light industrial trucking. It is not suitable for situations where more severe wear occurs and it is therefore chiefly used in hospitals, schools, garages, laboratories, cigarette and confectionery factories and breweries[4,5]. Some examples that typify the use of self-levelling epoxy floors are given below.

A food factory required a noiseless floor with non-slip characteristics. An area of 2500 m^2 of an epoxy system was laid in work rooms, loading bays and refrigerators and the floor was in good condition after several years' wear.

A car showroom required a floor which was not only attractive in appearance but also easy to clean and resistant to oil. An epoxy floor was installed and remained completely intact after 5 years' use.

A self-levelling epoxy floor was laid in the bottling plant of a brewery to provide protection against oil, grease and detergents together with mechanical wear.

The factory of a jam manufacturer had a suspended floor which suffered from considerable attack from fruit juice and sugar syrup and the original concrete floor was so badly attacked that liquids were percolating through to the rooms below. It was therefore essential to provide a topping that would be resistant to chemicals and completely impervious to water. It also had to be ready for use not more than 3 days after shutdown of the factory. A self-levelling epoxy floor was installed and production was resumed after 36 h.

A workshop producing components for the car industry suffered from oil spillage and wheeled traffic. Over the years, the floor surface had seriously deteriorated and the necessary repairs could be carried out only during holidays or shutdown periods. It was therefore decided that only an epoxy floor would meet the performance and time criteria. The old floor was cleaned and scarified and holes filled with an epoxy mortar. A primer was applied, followed by the self-levelling epoxy system. Ducts and gullies were repaired by cutting out defective areas and filling with an epoxy composition. This example demonstrates the advantages of epoxy floors in remedial work, since hardening occurred after 2 or 3 h and the floor could withstand foot traffic after 12–24 h, full chemical resistance being obtained in a few days.

Trowelled floors

Most epoxy resin used in flooring is in the form of trowelled industrial floors. These are filled compositions that are applied by trowel over a prepared sub-floor to a thickness of 1/8–1/4 in (about 6 mm) and provide a skid- and wear-resistant, chemically impervious topping. The binder constituents are usually arranged so that a simple mixing ratio of resin to curing agent components is achieved, such as 1:1 by volume. When the binder has been mixed, usually be mechanical means, the aggregate is added and mixing continued. Experience has shown that it is essential to add the aggregate after the binder constituents have been mixed if thorough mixing of all ingredients is to be achieved. Silica sand or calcined bauxite are often preferred aggregates, used at an aggregate:binder ratio of 7:1.

The choice of aggregate for a trowelled mix is as important as the choice of the resin and curing agent components. Important aggregate properties are as follows:

(1) Particle size and shape.
(2) Particle size distribution.
(3) Particle density.
(4) Particle hardness.
(5) Chemical inertness.
(6) Price.

After application, the floor will be sufficiently hard to withstand normal usage within 24 h and will attain full chemical resistance strength after 7 days. This speed of hardening compared with that of a cement material has led to these floors being used where it is important to minimise the time taken for floor repair and maintenance. An indication of the installed cost of a trowelled epoxy floor in the U.K. in 1972 was £3.00–5.00/m^2. Clearly, costs will vary from country to country and also between different applications. Surface preparation may be straightforward and minimal or lengthy and time consuming; there may be complicated skirtings and channels to be formed, and these may represent a high proportion of the total work to be carried out. These difficulties of estimating an accurate general cost are common to most systems and have not prevented trowelled epoxy floors being widely used in applications where good chemical resistance and durability are required and where speed of opening the floor to traffic is of great importance. Typical examples of some of these situations are given below:

(1) In chemical plant and in sugar and oil refineries where corrosion of concrete occurs due to attack by acids and alkalis.
(2) In plating and pickling shops and in tanneries and other factories where severe acid spillage and fumes occur.
(3) In food and soft drink factories, dairies and industrial kitchens where attack from acids is coupled with wear from steel-shod trucks and severe cleaning methods are used.
(4) In garages, engineering work shops and machine shops where oil and grease occur together with heavy wheeled traffic.

More specific examples of the use of trowelled epoxy floors are given below and will help to demonstrate the outstanding advantages of these floors.

The U.K. factory of the world's largest manufacturer of gloving leathers had 2400 m^2 of its tannery floor resurfaced using acid-resistant epoxy flooring ¼ in (6.3 mm) thick. This epoxy floor has reduced floor maintenance virtually to nil. During the tanning process, skins are pickled in sulphuric acid for temporary storage and during subsequent processes formic acid is used to lower the pH of the dye bath to 2.5–3, and the exhausted dye baths are washed out and drained on to the floor. Dyeing and pickling inevitably involve spillage; heavy trucks carry hides about the tannery. All these processes lead to the rapid disintegration of concrete floors. It was therefore decided to use a trowelled epoxy floor in order completely to protect the concrete floor, the cost of the epoxy being only half that of traditional acid-resistant floors such as quarry tiles which were used in the past[6].

A brewery floor in Manchester was required to be waterproof, non-dusting, easily maintained with an attractive appearance, have non-slip characteristics when wet or dry and withstand foot and vehicular traffic and chemical attack over a wide temperature range. In particular, it would be subjected to sugar, acids, yeast and beer spillage and caustic soda in the temperature range 10–77 °C together with wear from heavy metal-banded casks and heavy loaded steel-wheeled trucks. The trowelled epoxy floor was easily laid during the 12-day shutdown period for the brewery and has proved itself to be virtually maintenance free and to meet all of the other performance requirements that were specified.

A suspended floor in a factory was subjected to oil saturation and constant iron-wheeled traffic. In order to prevent complete breakdown of the floor and danger to the workers on the floor below, it was necessary either completely to re-lay the concrete or to provide a protective topping. The latter course was adopted and a trowelled epoxy floor was installed that proved to be a far cheaper alternative than re-laying the concrete floor.

Epoxy terrazzo[7,8]

Over the centuries, conventional terrazzo floors have been laid to provide a decorative, hard-wearing floor surface. This type of floor consists of marble aggregate held in a cement matrix and is normally laid on a 1–2 in (25–50 mm) sand base. After the cement has hardened sufficiently, which may take perhaps 7 days, the surface of the floor is ground so as to expose the aggregate and to smooth the surface. Epoxy resins can replace the Portland cement matrix with the following advantages:

(1) The epoxy floor will bond to any substrate and possesses high tensile compressive and flexural strength. It can therefore be laid in thin layers, which allows a saving in weight to be made compared with the cement terrazzo. Moreover, there is no need for a sand bed to be used, which again saves weight and decreases the overall floor thickness.

(2) In 2 days, the epoxy terrazzo can be installed and ready for use. The cement terrazzo takes at least 7 days before it can be used.
(3) The epoxy system can be easily formed into skirts and fillets and laid around columns, staircases, equipment, etc. No zinc or brass divider strips are needed, which allows a cost saving to be made.
(4) The performance of the epoxy floor is superior to that of the cement terrazzo in terms of flexibility and impact resistance, mechanical strength properties and adhesion and the ability to withstand corrosive materials. Both types of floors should give comparable service under normal conditions.

The application of epoxy terrazzo is very similar to that of heavy-duty trowelled floors. The substrate must be sound, clean and dry. A tack coat of binder is then applied, followed by the epoxy binder–aggregate mixture. After trowelling to form a floor about ¼ in (6 mm) thick, the resin is allowed to harden. As soon as possible, the surface is then ground smooth using water as a lubricant. Any pin holes that appear in the surface should be filled with an epoxy grout and the surface re-ground. Finally, the surface can be polished if a very smooth finish is required.

Typical locations where epoxy terrazzo has been installed include food factories, laboratories, banks and office foyers.

FUTURE DEVELOPMENTS IN EPOXY FLOORS

In order to ensure that epoxy floors maintain their high growth rate and continued penetration of the quality industrial flooring market, improvements will need to be made in the following areas:

(1) Epoxy floors will need to be made resistant to steam cleaning. At present they are failing, possibly due to the difference in the coefficients of thermal expansion of the floor and sub-floor concrete, or to the increase in the vapour pressure of the moisture held in the concrete sub-floor.
(2) Sand-filled floors often wear excessively if they are subjected to wet trucking, probably owing to sand worn from the surface forming a 'grinding paste' and causing rapid wear.
(3) For the efficient laying of trowelled floors, skilled labour is required. A mechanical means of laying would speed up the process and decrease the necessity for skilled manual trowelling.

Typical laying speeds (m^2/man day) for various types of floor are:

Plastic tiles	50
Linoleum	200
Ceramic tiles	20–25
Epoxy trowelling	9–10
Epoxy self-levelling	50–70
Concrete (3–4 cm)	40–50
Wood parquet	40

The speed of laying trowelled epoxy flooring needs at least to be doubled, but any equipment developed to achieve this speed should be compact and portable. Complex machinery would not normally be acceptable for the typical sizes of floors (1000–3000 m^2) for which epoxy resins are used.

REFERENCES

1. *Flooring,* **13**, No. 1, 4 (1968)
2. STANLEY, T. J., *Flooring,* **14**, No. 9, 2 (1969)
3. LILLEY, A. A., *Civ. Engng. publ. Wks Rev.,* **62**, No. 737, 1375 (1967)
4. *Flooring,* **12**, No. 2, 26 (1967)
5. *Fd Mf,* **41**, No. 12, 41 (1966)
6. *Flooring,* **13**, No. 1, 22 (1968)
7. *Epikote Resin Terrazzo Floor and Wall Surfaces,* Shell Chemical Company
8. *Bldg Constr.,* **33**, No. 3, 58 (1963)

PLATE 8
An epoxy-based mortar was used for jointing the superstructure of the Parc des Princes stadium in Paris (Courtesy of CIBA-Geigy)

PLATE 9
An epoxy resin floor in a heavy electrical equipment factory (Courtesy of Shell)

12

Roads and Bridges

EXPERIENCE IN THE U.S.A.

The first large-scale trials of epoxy resin road surfacing materials were carried out in the U.S.A. in the mid-1950s[1,2]. The idea behind this use of the resins was to utilise their excellent adhesive properties to bond sharp, hard aggregate on to road surfaces that had become smooth through traffic wear and thus to improve the skid resistance of the surface. Other additional advantages of an epoxy surface on roads were quickly identified: (a) the epoxy layer was an effective barrier to moisture and moisture vapour and (b) the chemical resistance of the resin surface layer protected the underlying road material from chemical attack by oils, grease and de-icing chemicals. The latter advantage was realised to be of special value on bridges[3] and the protective aspect of the epoxy surface eventually became more important than the original aim of improving anti-skid properties[4]. Incidentally, the U.S.A. has a greater proportion of concrete bridges than most other countries.
By 1960, most U.S. State Highway Authorities were evaluating epoxy membranes as protective surfaces on bridges. Despite the high cost of the materials involved with epoxy systems, it was realised once the tests were underway that epoxy surfaces would far outlast conventional cement or asphalt surfaces. This was especially true in situations with heavy traffic, such as bridges and toll-booths. Epoxy resins gave a clear cost advantage by reducing the frequency of replacing the road surface and removing the need for raising drain covers, kerbs, etc., which is necessary when re-laying with a conventional material which needs to be put down in thick layers. A further important advantage was the lower weight of a resin surface compared with that of surfaces of other widely used materials[5].

The epoxy system used in this work consisted simply of a liquid resin extended with coal tar and hardened with diethylenetriamine (DTA). A flexibiliser, such as pine oil, was also frequently included and it also served to improve the compatibility between the various components of the formulation. The road surface to be treated was first cleaned by washing and acid etching where necessary and the epoxy system was sprayed on to the road surface using a truck-mounted spray bar served by positive metering pumps and a mixing head. The rate of application of the resin mixture was about 2–3 lb/m^2 (1 kg/m^2). While the resin was still soft, sharp clean sand or other aggregate was broadcast over the surface, the excess being swept away with brooms after the resin had fully hardened. The total thickness of the resulting overlay was about 1/8–1/4 in (3–6 mm). There are many accounts of the large-scale application of epoxy surfaces to bridges that were carried out in the U.S.A. in the early 1960s. The protective non-skid surface was applied to new bridges such as the Intercity Viaduct in Kansas City[6,7], where 54 000 m^2 were laid, but remedial work on existing bridges represented a far wider use. The San Francisco Oakland Bay bridge[8] surfaced with 150 000 yd^2, the Triborough bridge, New York[9] with 160 000 yd^2 and the Dallas bridge over the Columbia River on the Oregon–Washington State Line[7,10] with 10 500 yd^2 are but three examples of this very large application of epoxy surfacing to protect existing bridge structures.

A massive investment of \$1M was begun in 1963 by the New Jersey Turnpike Authority to seal all existing and new bridge surfaces with an epoxy–coal tar system[11–14]. It has been estimated[15] that up to 1966 some \$10M had been spent on epoxy surfacing by U.S. Highway Authorities.

Not all of the applications of epoxy surfaces have been successful. Some have failed owing to lack of adequate surface preparation, leading to loss of adhesion, and others because the use of large aggregate led to capillary formation which formed weak points and ultimately holes in the epoxy membrane[15]. Experience has also shown that chains fitted to tyres in winter weather have caused the breakdown of a number of surfaces. The Triborough Bridge, New York, and the New Jersey Turnpike bridges have all suffered in this respect. Where chains are not normally used in winter, the surfaces have mostly remained in excellent condition.

EXPERIENCE IN EUROPE

Up to the mid-1960s, very little surfacing work of this type had been carried on outside the U.S.A. Many small trial patches had been placed in the U.K., but there was a reluctance to accept the U.S. trial results as valid for U.K. conditions and it was felt that the results of the U.K. testing programme should be awaited before a final assessment and firm recommendations could be made[16]. A similar situation also existed in other European countries. Moreover, contrary to the situation in the U.S., there were very few bridges in the U.K. that did not have a protective bituminous surface on top of the concrete slab and, in addition, the use of de-icing chemicals was far less and winters less severe than in the U.S.A. It now appears, however, that the long-held assumption that a properly applied bitumen bridge surface is impervious to water is incorrect. There is some evidence that in the long term bitumen surfacings can become slightly porous owing to micro-cracking. This cracking would be exacerbated by severe winter conditions, such as exist in Scandinavia. Once moisture has penetrated through the bitumen to the concrete beneath, the progressive break-down of the concrete begins and the mechanical effect of freeze–thaw cycling is particularly severe on concrete. Cracks in the concrete lead to spalling and ultimately to failure. It has been estimated[12] that the cost of patching and repairing concrete bridges on the New Jersey Turnpike alone between 1958 and 1965 was $6M.

The realisation that water might pass through a bitumen surface led to the use of an epoxy system purely as a water-proofing layer between the structural concrete of a bridge and the wearing surface of bitumen or modified epoxy resin. Thus, in 1966 a bridge on the New Jersey Turnpike had an impermeable epoxy resin membrane applied to the concrete bridge deck and a wearing surface of bitumen was placed on top. A similar technique has been used in recent years on many bridges on the Autostrada in Northern Italy, some 300 000 m^2 of the bridge surface being treated by 1970. The general method adopted in these cases was to apply an epoxy–coal tar system to the patched and cleaned concrete base. The resin layer was then dressed with sand, care being taken to use rounded sand particles so as to avoid the effect of capillary formation with flat particles. The particle size distribution was also found to be important in ensuring the presence of sufficient spaces between particles, so as to obtain a continuous film of the resin. The fittings on the bridges were then raised 1½ in (38.1 mm) to contain the final wearing surface, which was a layer of neoprene-modified asphalt with asbestos added.

The extreme temperatures that occur in Sweden led, in 1966, to the Gothenburg harbour bridge being treated with a similar multi-layer method[17]. The epoxy system, laid on to the prepared concrete bridge deck, served as a sealing coat and adhesive for a neoprene rubber layer. This layer was applied before the epoxy was fully cured, but was sufficiently hard to allow equipment to be placed on it. Lastly, after the neoprene had hardened, the third layer, consisting of an epoxy–coal tar system, was applied and the surface dressed with fine sand. This multi-layer membrane technique may well be fully justifiable under Scandinavian conditions, but would probably be uneconomic in many other countries where the sandwich method as used on the Italian Autostrada bridges would be preferred. In all of these applications, the epoxy system must have good adhesion to the concrete, be impermeable to water and de-icing chemicals and also be flexible. This flexibility is required in order to accommodate the relative movement of the topping and its substrate and also to ensure that the membrane remains as a continuous film when hairline cracks in the concrete are subject to movement.

Many European countries, including the U.K., have a large number of iron bridges, many of which are in railway systems and which frequently have concrete decks. They also present a particular corrosion problem as not only are protective coatings required for the structural steelwork, but also the bridge deck needs to be protected. Epoxy-based products are being used as impermeable layers on these decks.

LONDON ANTI-SKID PROGRAMME

From the above discussion, it can be seen that in the U.S.A. the original purpose for using an epoxy resin system to obtain non-skid road surfaces gave way, in the mid-1960s, to the use of epoxy resins primarily for protection against water and chemicals. At about the same time, consideration was being given in London to the use of epoxy resin systems as a non-skid treatment for urban roads. Each year in the London Metropolitan Police District, 50 000–60 000 accidents occur which result in personal injury or death. The average cost of an urban injury accident was estimated to be £880 in 1967[18], which at 1974 prices is about £1600. The total cost to the community of road accidents in London alone therefore amounts to £70–£84M per annum.

A study of these accidents in London showed that 70 per cent occurred at or within 20 m of a road junction. Skid tests also showed that the general level of skid resistance was reduced in the vicinity of road junctions, pedestrian crossings, roundabouts, etc., and these are also the areas that are most subject to polishing by vehicle wheels while braking, turning or accelerating. It was therefore decided to investigate ways of improving the skid-resistant properties of road surfaces.

Burning off the original asphalt and resurfacing with road asphalt as a means of improving skid resistance was rejected because it was found that the skid resistance of the new surface quickly deteriorated. Mechanical roughening was carried out on trial areas but the improvements obtained were short-lived. Finally, surface dressing was considered and, as many of the areas to be treated were subject to very severe traffic conditions, a synthetic resin binder was felt to be necessary.

In 1966, small-scale trials were carried out in Central London which led to an epoxy–bitumen extended system being chosen by the Greater London Council (GLC) for major full-scale trials at seven important road junctions where a number of accidents involving skidding had occurred. These trials were carried out in 1967 and two sites suffered early partial failures caused by a fault in the mixer of the application machine. At the remaining five sites, there was an immediate dramatic reduction in the number of accidents. After 1 year, the five sites showed an overall reduction of 50 per cent in the number of accidents compared with the previous year, after allowing for the changes in the pattern of accidents in London as a whole during that time. These results encouraged the GLC to expand this programme of road treatment in the autumn and winter of 1967–68. Nineteen sites were treated, including prominent intersections in Central London such as the Strand, Oxford Street and Charing Cross Road. During 1969–70, a further 72 sites were treated and other large programmes were carried out in 1970–71, 1971–72 and 1972–73.

In most of these applications, 'Shellgrip'[19,20], a modified epoxy system developed by Shell, was used. This material is applied hot at a minimum rate of 2½ lb/yd^2 (1.35 kg/m^2) on to the road surface, which has first been swept clean with brooms. No other surface preparation is required apart from the removal of any very oily patches. The recommended aggregate, calcined bauxite, is then spread over the surface at a rate of about 12 lb/yd^2 (6.5 kg/m^2). This aggregate is more resistant to polishing than any known natural road stone and maintains its skid resistance at high levels for a very long time. The resin is applied using special truck-mounted automatic spraying equipment, which heats the two components, meters and mixes them together and applies the mixture through a spray bar on to the road surface. The aggregate is applied mechanically with a grit spreader and the excess is recovered with automatic broom or vacuum equipment. The resulting rate of application of aggregate after removing the excess is about 9 lb/yd^2 (5 kg/m^2). The time required for the resin film to harden sufficiently to allow traffic on to the newly surfaced road depends on the ambient temperature and varies from 2 h at 20 °C to about 5 h at 5 °C[21]. It can be estimated that the approximate cost of treating a site (1973) is £1.70–£2.00 per square yard, the average site size being 1400 yd^2. The work is usually carried out overnight so as to avoid traffic congestion and the treated road must be open for the morning commuter traffic.

Full accounts of this use of epoxy resins in London were given by Hatherley and co-workers[21,22] and *Table 12.1*, taken from the paper

Table 12.1

ACCIDENT STUDIES MADE ON LONDON ROAD JUNCTIONS (FROM HATHERLEY AND LAMB[21])

41 sites were treated with epoxy resin and calcined bauxite.
Summary of results from October, 1968, to the end of September, 1969.

Accidents	*Before treatment*	*After treatment*	*Change, per cent**
Total accidents	288	200	–31
Total casualties	337	258	–23
Accidents on wet roads	105	29	–72
Accidents on dry roads	183	171	–7
Rear-end collisions	37	10	–73
Loss of control	23	7	–70
Crossing collisions	54	31	–43
Turning collisions	23	18	–22
Pedestrian accidents	76	65	–14
Other accidents	75	69	–8

*During the same period, accidents in the Metropolitan Police District as a whole increased by 1.5 per cent.

by Hatherley and Lamb[21], shows a study of accidents at 41 road junctions before and after treatment with epoxy resin. The dramatic reduction in accidents, especially on wet roads, it immediately apparent. Total accidents were reduced by 31 per cent and total casualties by

Table 12.2

POSSIBLE 10-YEAR PROGRAMMES (FROM HATHERLEY AND LAMB[21])

Forecasted accident and cost reductions on Metropolitan road junctions using resin–bauxite treatment (1970 values)

Initial assumptions:		
	Cost per site	= £2000
	Accident reduction	= 30 per cent
	Cost of personal injury accident	= £1090
	Mean life of treatment	= 5 years
	Total number of accidents on all roads in London area	= 56 000 p.a.

Annual expenditures per annum for 10 years	*£100 000*	*£200 000*	*£300 000*	*£400 000*	*£500 000*	*£600 000*
No. of sites treated per annum (including maintenance)	50	100	150	200	250	300
Total No. of sites treated after 10 years	375	750	1125	1500	1875	2250
Total accident reduction in 10-year period	10 800	17 300	22 300	26 300	29 600	32 700
Total cost of accidents saved in 10-year period	£11.8M	£18.8M	£24.3M	£28.7M	£32.2M	£35.6M
Total cost of anti-skid treatment	£1M	£2M	£3M	£4M	£5M	£6M
Cost of preventing one accident in 10th and subsequent years	£61	£78	£92	£107	£119	£131

Note: The annual expenditure after 10 years would continue at about the same rate but without increasing benefit. The continuing expenditure would be required to maintain the treated areas.

23 per cent. On wet roads there was a 72 per cent reduction in the total number of accidents compared with only a 7 per cent reduction on dry roads. During the same period as the study (October, 1968–September, 1969), accidents in the London Metropolitan Police District as a whole increased by 1.5 per cent. Hatherley and Lamb[21] also considered the economic case for a number of possible 10-year programmes and these are given in *Table 12.2*. Assuming a total accident reduction of 30 per cent and using other reasonable assumptions for the cost of a personal injury accident, the total number of accidents on London roads, and the cost of anti-skid treatment, Hatherley and Lamb forecast a total saving on the cost of accidents over 10 years of £11.8M for an outlay of £100 000 per year for 10 years rising to a saving of £35.6M over the 10-year period if £600 000 per annum (£6M in total) were spent on anti-skid surfacing. While some of the assumptions used and their values may be questioned, the experience in London is surely sufficient to warrant all urban traffic and road safety authorities in all countries examining the possible application of epoxy anti-skid surfaces to their own roads subject to heavy traffic.

EXPANSION GAP NOSINGS FOR BRIDGE DECKS

Experience is showing that high-speed, high-density traffic on motorways is causing rapid wear and deterioration of the asphaltic or concrete wearing surfaces about 6 in (15 cm) on each side of the expansion gaps. Rapid wear has been particularly noticeable on bridge decks where vertical movement can occur between adjacent sections.

In the past, expensive steel plates have been used but difficulties have been experienced in fastening these permanently to the bridge deck. After extensive development work by the U.K. Ministry of Transport[23], an epoxy resin mortar has been produced which is eminently suitable for providing an economic and durable method of preventing wear at the edges of expansion joints. Bridge expansion joints with epoxy mortar nosings have the advantages over bridges with steel nosings that they cost 50 per cent less and they are quicker and easier to install and replace. These nosing compositions are now specified for the construction of expansion joints in most bridges and road projects in the U.K.

FUTURE DEVELOPMENTS

The well documented use of epoxy anti-skid road surfaces in London will no doubt lead to many other cities and towns taking up this development. Already (1973) a number of U.K. cities have trials under way, and in many other European countries, such as Belgium, Germany, Italy and Sweden, many applications of epoxy systems are already proving their worth. The treatment of bridges will also develop further. In the U.K. the surface of the slow lane of the Severn Bridge, constructed in 1966, has been given an epoxy anti-skid surface in order to prevent skidding accidents that would block the bridge[24].

Future technical developments that might occur in this surfacing work are the replacement of the calcined bauxite with a cheaper but equally effective aggregate and improvements in the resin binder system. Trials are already under way to investigate polyurethane binders, which would be expected to be cheaper and harden more rapidly than epoxy resins and hence allow a larger area of road surface to be covered in a given time. However, there are certain limitations to the use of these resins, such as their water sensitivity and toxicity.

REFERENCES

1. NAGIN, H. S., NOCK, T. G. and WITTENWYLER, C. V., *Bull. Highw. Res. Bd, Washington D.C.*, No. 184, Paper 1 (1958)
2. NAGIN, H. S., HERR, D. S. and NOCK, T. G., *Bull. Highw. Res. Bd., Washington D.C.*, No. 188, Paper 13 (1958)
3. Symposium on Effect of Deicing Chemicals on Structures. *Bull. Highw. Res. Bd*, Washington D.C., No. 323 (1962)
4. WITTENWYLER, C. V., *Highw. Res. Abstr.*, **32**, 7 (1962)
5. LANGE, H. F., *Publ. Wks, N.Y.* **96**, No. 6, 113 (1965)
6. *Amer. City Mag.*, 79, No. 1 (1964)
7. MINARIK, W. L., *Publ. Wks, N.Y.*, **94**, No. 7, 97 (1963)
8. *Contractors Engrs Mon.*, **61**, No. 9, 58 (1964)
9. *Contractors Engrs Mon.*, **60**, No. 9, 22 (1963)
10. MACKENZIE, T. T., *Civ. Engng*, **32**, No. 3, 42 (1962)
11. RILEY, O., *Publ. Wks, N.Y.*, **94**, No. 5, 126 (1963)
12. RILEY, O., *Publ. Wks, N.Y.*, **97**, No. 6, 106 and 164 (1966)
13. FISHER, R. L., *Publ. Wks, N.Y.*, **96**, No. 5, 100 (1965)
14. *Rural Urban Rds*, **3**, No. 7, 58 and 85 (1965)
15. McCONNELL, W. R., *Symposium on 'Epoxies with Concrete'*, Publication SP-21 Paper SP-21-3, American Concrete Institute, Detroit, U.S.A. (1968)
16. JAMES, J. G., *Rds Rd Constr.*, **41**, No. 488, 236 (1963)
17. *Epikote Resin Based Impermeable Membranes for Concrete Bridge Decks*, Shell Chemical Company

18. DAWSON, R. F., *Rd Res. Rep.* No. LR-79, Ministry of Transport, London (1967)
19. *Safer Roads with the Shellgrip System,* Shell International Petroleum Company
20. SHELL INTERNATIONAL PETROLEUM COMPANY, Brit. Pat., 1, 086, 390
21. HATHERLEY, L. W. and LAMB, D. R., Paper presented to 6th World Highway Conf., Montreal (Oct. 1970)
22. HATHERLEY, L. W., MAHAFFY, J. R. and TWEDDLE, A., *J. Instn. Highw. Engrs,* **16**, No. 4, 3 (1969)
23. *Technical Memorandum (Bridges),* No. BE6, Ministry of Transport, London (1967)
24. *Financial Times, London,* p.6, 8th November 1971

13

Adhesion and Grouting in Building

A significant increase in the use of epoxy adhesives by the civil engineering industry has taken place over the last 10 years. This has been chiefly due to two factors:

(1) The increasing realisation that the bond strengths of epoxy adhesives are considerably greater than the cohesive strength of concrete, which is the most common load-bearing construction material used in civil engineering. The tensile strength of good quality concrete varies between 250 and 700 lbf/in^2 (1.75–5 MN/m^2) according to the mix used, whereas the tensile strength of epoxy adhesives exceeds 700 lbf/in^2 (5 MN/m^2) and is often much greater, up to 8000 lbf/in^2 (56 MN/m^2).

(2) The ability to make significant savings in time by using an adhesive. The rate of development of mechanical strength of epoxy adhesives is therefore important in this context, and it can vary widely depending upon the epoxy system used, the ambient temperature and other factors. However, it is certainly considerably faster than the rate at which concrete develops comparable proportions of its final strength.

The uses for epoxy adhesives and grouts can be classified broadly into:

(a) Remedial work, such as strengthening and repair of existing structures.

(b) New work, where the use of adhesive was envisaged at the design stage.

Most of the early uses of epoxy adhesives in the construction field were concerned with the repair and conservation of structures. The specification of epoxy adhesives in the design of new structures came later, when much more was known about the behaviour of the materials and consequently more confidence in their use was developed. But even now the proportion of new structures in which epoxy adhesives are incorporated is relatively small and the use of the adhesives tends to be confined to stress-distributing layers in the construction of post-stressed structures composed of prefabricated units.

In contrast, there is now a large number of widely differing examples of the remedial use of epoxy systems in both new and old constructions. Typically, the adhesives and grouts are used to repair structures that have begun to deteriorate in service owing to concrete spalling, concrete cracking, overloading, etc., or to repair damaged prefabricated units, or to make good minor design errors and faulty workmanship that have become evident on site.

REMEDIAL USES

Concrete crack repair

Most conservation work in which epoxy adhesives and grouts are used is concerned with concrete structures that have become degraded. Cast concrete has a tendency to develop hair cracks after curing or ageing or can become cracked as a result of bad design, the application of unforeseen stresses, bad workmanship or external damage. Hair cracks, although not serious in themselves, can join together to form a continuous network of cracks, which can then be enlarged either by movement within the structure or by the freeze–thaw action of water. Large fissures can develop and spalling of the concrete can occur. This effect has been observed in buildings, roads, bridges, dams, breakwaters, etc., and it is essential to fill the cracks and restore the structure before excessive damage occurs.

Epoxy resin grouts are widely used to seal these cracks, restore the integrity of the structure and prevent the ingress of water, which would cause spalling and corrode any metal reinforcement. The adhesive and cohesive strengths of the epoxy grout are greater than the cohesive strength of the concrete and these strengths are maintained after long-term ageing and exposure to the alkaline effect of cement and concrete. Typical strengths are:

	Compressive strength	*Tensile strength*
Epoxy grouting compound	14 000 lbf/in^2 (98 MN/m^2)	7000 lbf/in^2 (49 MN/m^2)
Cement concrete	5000 lbf/in^2 (35 MN/m^2)	500 lbf/in^2 (3.5 MN/m^2)

Other valuable properties of the epoxy grouts are:

(1) Their good dimensional stability and low shrinkage compared with cementitious grouts.
(2) Their rapid rate of development of mechanical properties. The epoxy systems attain full compressive strength in 24 h, whereas cement requires 7 days or longer. This is very important if the down-time for a machine is costly, e.g. perhaps up to £1000 per day, or if a road or bridge has to remain closed. The cost of raw materials in these situations becomes unimportant compared with the value of the time saved.

The alternative to using an epoxy grout is usually either to use a cementitious grout or to dismantle the construction, if that is possible, and to reconstruct. Clearly, there are a number of situations, such as in dams and bridges, where grouting is the only practicable solution and in these instances epoxy compositions are rapidly becoming the preferred system. However, before deciding upon the use of an epoxy grout, it is important to identify the cause of the cracking. If the structure is being subjected to stresses beyond the design maximum, cracking will probably continue to occur even after the epoxy grout has been used. Transient cracking can be overcome but it is doubtful whether cracks due to inherent overstressing can. Similarly, the repair of prestressed concrete that has broken down will not restore the original strength of the member.

The practical methods used to fill cracks in concrete or repair spalled areas have been fully described elsewhere[1-4]. One technique consists in cutting out major cracks to a depth of about 6 mm, removing dust and loose masonry and bonding injection nipples along the length of the cracks. In another technique, advocated by Gaul and Smith[2] as being faster, a sealer is merely painted over the cracks, leaving unsealed areas to serve as entry ports. The resin is injected first at the lowest port and when the resin appears at the next highest port, the lower port is sealed. Injection is then continued at the second port. Gaul and Smith[2] reported that this process is applicable to cracks wider than 0.005 in that can be cleaned and under favourable conditions, cracks as narrow as 0.002 in have been successfully filled to depths of 60 in and lengths greater than 100 ft. A review of the techniques of applying epoxy grouts was given by Tsurata[2].

There are very many examples of the successful use of this method to repair widely differing types of structures, such as concrete pipes, crane ways, support beams, cracked floors, bridge piers, building walls, dams, harbours and wharf piers. Many of these case histories have been reported in the literature and also in the publication "Construction", issued by the Adhesive Engineering Company, California, U.S.A., which describes applications of its widely used epoxy grouting system. The following is a selection of the many case histories, which it is hoped will demonstrate the wide versatility of the epoxy grouting systems and the financial advantages to be gained by using them.

One of the earliest uses of epoxy grouts was in the repair of a wharf at the Port of Los Angeles in 1962[5]. Large structural cracks had appeared in a number of concrete pile caps, and the alternative to filling the cracks with epoxy grout was to demolish the entire wharf deck and to reconstruct the beams and deck with new concrete. It was estimated that the repair with epoxy grout would cost about a quarter as much as reconstruction. In addition, rebuilding would result in a large loss in revenue while the port was out of use, whereas crack repair could be made without interrupting shipping operations and would lead to a rapid restoration of the wharf to full service. It was decided to use an epoxy grout and some 5000 linear feet of cracks were repaired, which required nearly 3 tons of grout. The total cost of injection, including all preliminary cleaning and shot-blasting was \$93 000; the estimated cost of replacing the whole wharf was \$350 000 and the minimum estimated loss in port revenue, assuming that the rebuilding took 1 year to complete, was \$100 000. Hence the apparent saving by using epoxy grout was \$357 000.

A further example of dockside repair was at the site of a bulk-loading terminal in Vancouver, British Columbia. Six pre-cast, pre-stressed piles were cracked and were repaired in 6 days using an epoxy grout, at a cost of \$1800. The cost for replacement of the piles would have been \$18 000[6].

As examples of the repair of dam walls, at the Yuta Dam in Japan, the Hauser Dam in Montana and the Twin Lakes Dam in California extensive repairs were made to cracks with epoxy grouts. Many other examples of such repairs can be quoted, for example the Kansaki road bridge near Osaka, Japan, a railway bridge over the Salinas River, California, a road bridge over a Michigan State highway, the coke oven batteries of the Royal Dutch Hoogovens at Ymuiden, The Netherlands, and the 4-mile concrete ring of the U.S. National Accelerator at Batavia, Illinois.

In addition Gaul and Smith[3] quoted the following brief case histories, which illustrate the financial savings that epoxy grouts can achieve.

At a U.S. steel plant a 7-ton steel coil had fallen on to a 12 in reinforced concrete floor slab located over an air plenum. The floor was repaired *in situ* by means of an epoxy grout. Complete removal and replacement of the shattered concrete would have cost ten times more than the cost of the grouting method.

A reinforced concrete crane track had become cracked through overloading at a refinery of the Standard Oil Company of California. The cracks were repaired by injecting an epoxy grout, without interrupting plant operations. This method saved $16 000 over the alternative of dismantling the entire system and completely replacing the concrete. Eight years after the repair had been made there were no new cracks observed in the concrete and the repaired cracks had not enlarged.

The slab on the second floor of a building purchased by International Harvesters had deteriorated badly owing to overloading and the load-carrying capacity of the floor was reduced to 40 per cent of its original capacity. The replacement of the floor would have required the plant to be shut down and would have cost a minimum of $250 000. After careful trials and tests, it was decided to repair the cracked slab with an epoxy grout. Over a 4-month period, 30 000 linear feet were injected at a total cost of less than $100 000. There was no plant shut-down at any time during the operation and the repair was estimated to have restored the strength of the floor to 80 per cent of its original value.

The Alaskan earthquake in 1964 gave an opportunity for epoxy crack repairs to be used on a vast scale[7]. In many other countries also there are many examples that can be quoted[2,4]. Thus, the use of epoxy grouts in Poland has been described by Ciesielski[2], in Canada by Smith, Chognacki and Langhammer[2], and in Sweden by Fritzell and Warris[2].

Bonding concrete to concrete

Wide cracks or large spalled areas are often encountered in remedial work which would be uneconomic to repair with a filled epoxy mortar. These cavities are therefore either filled with a common cement-based mortar or the original piece of concrete that has broken away is bonded back into place. In both of these methods epoxy adhesives are used. In many instances, when wet uncured concrete is poured on to old mature concrete or steel, poor adhesion is obtained, especially when the new poured layer is relatively thin. An epoxy adhesive coated on to the old concrete surface, which must be clean, ensures good adhesion between the new and the old concrete. The adhesive, moreover, will cure satisfactorily while in contact with the wet fresh concrete.

The first epoxy adhesives to be used in this way were based on mixtures of the epoxy with polysulphide polymers, and systems of this type are still the most widely used today. These are two-component products consisting of the epoxy resin in one container and the poly-sulphide plus an amine curing agent in the other. The technique for using the adhesive is first to clean the old concrete surface by one of the various methods described in previous sections, and the two adhesive components are then mixed together and spread on to the clean surface. Before the adhesive is set, the new concrete is poured into place. The final bond strength between adhesive and the concrete surfaces is greater than the cohesive strength of the concrete itself. Typical situations where this bonding technique has been used include re-levelling after a structure has settled, making good poor concrete levels resulting from design errors and repairing large areas of damaged concrete such as severely pitted floors or road surfaces. Existing structures, such as dams, buttresses, pier supports and conduit walls, have also been extended by using an epoxy adhesive in this way and the resulting extended structure acts satisfactorily as a monolithic unit.

Examples of the use of the technique have been given by Gaul and Apton[8] and Shingleton[1]. In 1960, the Isewan typhoon badly smashed the large concrete sea-retaining wall along the coast at Ise Bay, Japan. The massive broken fragments of the wall were replaced in their original positions with shuttering around them. Their broken surfaces were covered with an epoxy adhesive, fresh concrete was poured into the gap between the two masses and the whole was allowed to cure. The resulting rebuilt wall was completely satisfactory, the adhesive giving an effective bond between the fresh concrete and both old concrete surfaces[9].

The impact strength of a thin layer of a cement topping on an old concrete base is usually very bad, but it can be improved by using an epoxy adhesive. This technique was used to resurface the 1400 m long concrete bridge over the river Ascheloos in Greece[10]. After repairing cracks in the structure, the bridge surface was cleaned, coated with the epoxy adhesive and the new concrete surface poured on top and levelled. Similar examples are the repair of 20 000 m^2 of floor in a

Paris building and the resurfacing of the Credit River bridge and parts of the Port Mann bridge in Canada[11].

Bonding reinforcements

Epoxy adhesives are extremely good at bonding dissimilar materials, and many examples exist in the construction field. It has been reported by L'Hermite and Bresson[2] and also by Kajfasz[2] that the flexural, bending and shear strengths of concrete beams can be greatly increased if sheet metal is bonded to the beam using an epoxy adhesive. This technique can be used where:

(1) The construction of a reinforced beam has been faulty.
(2) It is desired to increase the load-bearing capacity of a structure beyond its design limit.
(3) Damage to existing steel reinforcement needs to be repaired.

Almost all of the examples of the use of this technique come from South Africa. A bridge on the approach road to the Louis Botha Airport, Durban, was badly damaged when struck by the jib of a mobile crane. In order to effect a rapid repair with no interruption of traffic flow, the damaged concrete was removed and replaced with an epoxy resin mortar. A steel plate was then bonded to the underside of the patched-up beams so as to restore the bridge to its original design strength.

The same approach has been used to maintain numerous railway bridges along South Africa's coastline. Many of the bridges built in the 1920–30 period had deteriorated owing to severe spalling and to subsequent corrosion of the steel reinforcement, which had been exposed to the marine atmosphere. The bond strength between concrete and metal reinforcing rods can also be improved if the metal rods are first coated with an epoxy adhesive before pouring the concrete against the rods[12]; both resin and concrete cure at the same time. An alternative method is to coat the adhesive layer with sand before casting the concrete. It is claimed that this technique can improve the iron–concrete bond strength by 300–500 per cent. Metal reinforcing pins and rods can also be grouted in place by using an epoxy system rather than the usual cement-based grout.

Shingleton[1] quoted an interesting example in which steel reinforcement has been bonded in repairing and strengthening the Athlone bridge over the river Umgeni in Durban. Soon after the bridge was opened to traffic in May, 1968, major cracks appeared in the tips of the cantilevered supports that extend from the pillars of the bridge. Concrete beams rest on these cantilevers and form the bridge deck on which the traffic runs. After extensive testing, it was decided to restress each cantilever to its original design strength by bonding a stressing cable into the concrete with an epoxy adhesive. The tips were also reinforced by bonding a large number of steel rods into them with the same adhesive in order to prevent the concrete breaking when the cables were stressed. Concrete was then cast around the steel rods and any cracks formed when the new concrete cured were filled with an epoxy grout. All other cracks were filled in the same way and thus the structural integrity of the tips was ensured and the concrete was completely sealed against the ingress of water, which would corrode the steel reinforcement. A total length of more than 6000 m of holes was drilled in the bridge using diamond drills in order to bond an adequate number of steel rods.

EPOXY BONDING IN NEW STRUCTURES

There are a number of examples of the inclusion of epoxy adhesives in the design stage and in the erection of new massive structures, the adhesive mostly being used to join pre-cast concrete structural elements during erection. Small-section mullions up to massive bridge sections weighing perhaps 100 tons each have been bonded in this way.

The merits of building structural beams and arches from pre-cast units rather than casting in place are well known. Dry, non-bonded joints can be and are used, but unless they are specially keyed into each other and post-tensioned, they cannot adequately resist tension and shear stresses. Careful preparation and casting of adjacent elements is essential in order for accurate mating of the units to be possible. Upon this accuracy depend the proper uniform transmission of stress and the size of the gap between the elements, which, if open, will allow water and corrosive substances to penetrate the joint. There is therefore a need for an adhesive between the sections that will act as a load distributor and gap filler. Cement mortar joints at least 1 cm wide or an epoxy resin-filled joint can be used, but the resin adhesives have many advantages over the cement mortar and allow the maximum advantages of the pre-cast technique to be obtained. These advantages are:

(1) Faster cure and therefore props and supports can be removed earlier and post-tensioning begun sooner.
(2) Greater tensile and shear strengths are obtained, which can, in some instances, allow joined, untensioned units to be assembled before lifting into place as the joined sections are as strong as monolithic concrete.
(3) Shrinkage of the adhesive during cure is negligible and hence the joints remain watertight.
(4) Thinner joints can be used and are aesthetically more pleasing than the thick bands of cement mortar. Accurate precision casting of the concrete section is clearly necessary if thin joints are to be used. Alternatively, more time must be spent in grinding the mating surfaces to a close fit.

These advantages of using epoxy adhesives have been fully realised in the construction of a number of concrete bridges. Where a bridge is spanning a river or estuary or is in a situation where casting *in situ* is extremely difficult and therefore costly, a segmental construction technique is often used. In this method, pre-cast box sections are lowered successively into place as cantilevers from the bridge piers, high-strength epoxy adhesive being used as the jointing material.

Many large and complex bridges have been built in this way, including several notable bridges in France. The Pont D'Oleron links the island of Oleron, in the Bay of Biscay near the mouth of the Gironde, with the mainland. It is the longest bridge in France, stretching for 9928 ft (2900 m), and was erected in only 14 months as a continuous pre-stressed box beam from pre-cast concrete units bonded with an epoxy formulation[13-16]. The Paris ring-road system involved the building of two bridges across the Seine. One of these bridges, the Pont Aval, carries two carriageways, each of four lanes 3½ m wide, a central reservation 3 m wide, and has a total width of 34.6 m. Box beams of pre-stressed concrete form the deck of each half of the bridge and their weight varies from 40 to 70 tons each. The pre-fabricated box sections were cast against one another so as to ensure a perfect fit, the faces of each section being separated from each other solely by the epoxy composition. It was found that the use of the epoxy formulation enabled work to proceed independently of external conditions and ensured that all sections and the channels for the stressing cables were watertight[14,16].

Other bridges erected using this technique include the Pont Pierre-

Bénite near Lyon[13,16], a bridge on highway N186 at Choisy-Le-Roi near Paris[15,16], Casteljon bridge over the river Ebro, Pamplona, Spain[14], which required about 2 tons of adhesive, and the Rawcliffe bridge over the Dutch River in Yorkshire[14,17]. The segmental construction technique was also used in The Netherlands on three large motorway bridges[18,19], all associated with the development known as the Delta Project, which will lead to a huge extension of the industrial and residential area of Rotterdam and Europoort. The Kleinpolderplein fly-over is Holland's first three-level interchange and cost £7M to complete. It was constructed from 45-ton prefabricated box beam sections cast off-site and epoxy resin adhesives were used for the joints between the sections of the higher levels. The Brielse Maas bridge and the Hartel Kanaal bridge were also constructed by this technique and were epoxy jointed. But perhaps the success of the technique can best be demonstrated by the construction of the 5022 m long Ooster Schelde bridge, which was opened in 2½ years from driving the first pile and which required very little temporary scaffolding. Holland is in an advantageous position to use the pre-cast technique as the massive concrete sections can be transported to almost any part of the country by water.

Probably the most impressive road structure in which an epoxy jointing composition was used is the 1½ mile long elevated section of the N9 highway which follows an extremely sinuous route along the north-east shore of Lake Geneva between Chillon and Villeneuve in Switzerland (Plate 10). The road viaduct is mounted on slender piers, some of which are 135 ft (45 m) high, and pre-cast concrete sections, which weighed between 45 and 80 tons each, were erected by a giant travelling gantry crane that moved on rails attached to the bridge deck. The surfaces of all the pre-cast sections were coated with an epoxy composition before the elements were placed in position. When a pair of elements was provisionally placed, four 145-ton post-tensioning cables were fixed between the two elements. At a normal rate of fixing four pairs of elements per day, the viaduct was extended at over 300 m per month. Over 50 000 yd^3 of concrete and 5000 tons of steel were used in the 1½ mile viaduct, which cost £3M[14,20,21].

Epoxy adhesives have also been used to bond together large prefabricated concrete sections of buildings. In Durban, South Africa, a large sugar storage terminal was built using pre-cast sections, whereas most silos constructed previously were cast *in situ*. This latter method would have required the use of very expensive shuttering, which it would have been necessary to dismantle and move many times during the construction of the whole of the 800 ft (250 m) long structure. The building itself consisted of 80 principal arches, each leg of which was prepared from 18 cast units weighing 5 tons each. When erecting the units, the epoxy formulation was coated on to the joining surfacing and took up any lack of uniformity. After four arches had been completed, they were pre-stressed and the supports moved on to the next four to be erected[22].

One of the most exciting buildings architecturally to be built in recent years is the Sydney Opera House, designed by Jorn Utzon together with structural engineers Ove Arup and Partners. This cluster of shell forms was built from curved hollow concrete ribs, which themselves were made from pre-cast segments with matching faces. The whole string of segments was made continuous by post-tensioning across the transverse joints of the string. Each segment weighed about 10 tons and varied in width from 1 to 12 ft and in depth from 4 to 7 ft. An epoxy adhesive was used to bond the segments together and also served to maintain a uniform load distribution[2,22,23]. O'Brien[2] reported that the minimum and maximum compressive stresses that might be applied to the adhesive under extreme conditions would be 300–2500 p.s.i. (2–17.5 MN/m^2) with a maximum shear stress of 500 p.s.i. (3.5 MN/m^2). Under normal conditions the maximum compressive stress applied would be 1500 p.s.i. (10.5 MN/m^2) and the shear stress 100 p.s.i. (0.7 MN/m^2).

A further impressive use of an epoxy adhesive was in the construction of the superstructure of the new Parc des Princes Stadium in Paris, regarded as one of the world's most ambitious and successful structures in reinforced concrete (Plate 8). This structure, which has a capacity for 50 000 people, is composed of 50 huge arches of segmental concrete construction with internal steel stressing cables. An epoxy-based mortar was used for jointing between the hundreds of concrete segments, ensuring continuity of the structure and playing a fundamental part in load distribution both during construction and in the finished building. It also ensures that the stressing cables and their terminals are fully protected from the environment[24].

O'Brien[2] has also recorded many examples where slender columns and mullions have been joined with epoxy adhesives in buildings, such as Coventry Cathedral, the University of Exeter Science buildings, Somerville College, Oxford, and Abbotsinch Airport, Glasgow, and listed five types of joints that have been formed:

(1) Thin joints between pre-cast concrete units, which are post-tensioned after cure.
(2) Thin horizontal compression joints between sections of pre-cast concrete columns or mullions with no dowel connector bar.
(3) The same situation as in (2) but with dowels grouted-in using the same adhesive.
(4) Pre-cast concrete shear connections using bolts to keep the adhesive in compression.
(5) Various joints used in assembling pre-cast concrete staircases.

Nearly all of the joints considered so far have been of the type in which the adhesive is held in compression. There are not many applications known where the adhesives are used under conditions of shear stress, no doubt owing to the lack of long-term strength data on these adhesives, especially when used in thick sections and subjected to long-term stresses in various environments. However, studies have shown that epoxy resin adhesives can serve as reliable and safe shear connectors for composite T-beams under long-term static or dynamic loading[25]. In these tests, carried out on a number of composite T-beam designs, it was shown that the shear stress between the metal beam and the concrete slab seldom exceeded 200 p.s.i. (1.4 MN/m^2) in normal applications, whereas the ultimate stresses determined experimentally were between 600 and 1160 p.s.i. (4.2 and 8 MN/m^2), i.e. three to six times as great. None failed in dynamic loading where test cycling frequencies of 200–250 cycles per minute were being used. The shear stress in the joint during tests varied between 150 and 300 p.s.i. (1 and 2 MN/m^2).

Epoxy resin shear connectors would have an advantage over mechanical connections such as bolts by providing an improved distribution of the stress that may result from loading or shrinkage, impact or thermal cycling. Both techniques would cost about the same initially but in the long term the resin connectors would show savings on maintenance and repair.

There are a number of limitations to the use of epoxy adhesives that need to be overcome before they can be more widely used in construction, viz. fire resistance, bulk mechanical properties and creep.

Fire resistance

The epoxy compositions used in the applications considered so far cure at ambient temperatures. Their mechanical strengths decrease sharply at temperatures from 45 to 100 °C[26], depending on the curing agent used. Moreover, when the temperature reaches about 200°C, the polymer network starts to degrade and the adhesive breaks down physically. In comparison with concrete, therefore, epoxy compositions have a much lower temperature resistance, although epoxy systems may be formulated that approach concrete in resistance to flame penetration and heat transmission. Clearly, an epoxy-bonded stressed joint should not be used in a primary structure of a building where a fire risk is present unless the joint is pinned or post-tensioned, or unless it is a thin horizontal compression joint.

Bulk mechanical properties

In contrast to the thin bond lines achieved in metal bonding, the bond lines obtained with epoxy adhesives in the building field are several hundred times thicker. The bulk mechanical properties of the adhesive composition therefore become very important with respect to bond strength. Unfortunately, there are very few data published on the bulk mechanical properties of these adhesives at varying rates of testing. Some data were given by Shingleton[1] on the tensile and compressive strength of epoxy–polysulphide compositions which demonstrate the effect of bond thickness on bond failure for this particular system.

Creep

There are very few published data on the long-term strength properties of epoxy adhesives cured at ambient temperature. Johnson[26] described experiments with certain formulations and stated that with respect to one particular epoxy mortar 'it is suitable for use in structural joints in pre-cast concrete that transmit combined shear and compression and that its flexibility and creep are little greater than those of concrete'. The work showed the great importance of the correct formulation of the adhesive and in particular the need to exclude non-reactive diluents and plasticisers from the formulations. Clearly, much more systematic work is required on these properties before full confidence in the use of epoxy compositions in these situations can be developed.

MISCELLANEOUS BONDING APPLICATIONS

General-purpose epoxy adhesives are used in a wide variety of applications and to bond many different substrates. They can be used, for example, to bond concrete, stone and brick, glass and ceramics, vulcanised rubber, foamed plastics, cork, leather and textiles. The most important uses, however, are the bonding of light metals, iron and steel. The bonds are characterised by high mechanical strength and resistance to very many chemicals and solvents and, because perfectly sealed joints can be made, electrolytic corrosion between different metals can be avoided. Some examples that demonstrate the wide versatility of these products are given in the following paragraphs.

Cast concrete kerb-stones can be very effectively bonded to the road surface with an epoxy adhesive. Even more effective is the continuous extrusion on-site of kerbing from fresh non-slump concrete. This technique, which requires only two 1-in (2.5-cm) wide parallel strips of adhesive to bond 10-in (25-cm) extruded kerb, was described by Gaul and Apton[27] in 1960. At that time, the average U.S. National and Californian cost to install standard poured kerbs was about $2.50 per linear foot, reducing to $2.00 on long runs. The extruded method using epoxy bonding had a total cost for labour and materials of $1.00 per linear foot.

Road marker buttons and studs are also bonded to road surfaces using fast-setting epoxy compositions and motorway barrier supports have been glued into metal slots in the road and railway lines secured to concrete sleepers in a similar manner. Epoxy adhesives are used to carry out pile splicing on-site.

The extension of the 60-in (18-m) long piles of the Melbourne Cultural Centre in Australia is a good example of this technique[9]. The octagonal piles were driven almost completely into the ground and a steel collar placed round the last protruding 2 ft (61 cm). The collar was sealed at the bottom with a rubber gasket and 18 kg of an epoxy resin adhesive were poured into the collar. The next section of pile was lowered into the collar sleeve and thus into the epoxy mixture. This mixture was then cured very rapidly by surrounding the joint with a steam-heated collar. After 20 min, the collar was removed, the joint cooled and pile-driving recommenced.

Very many other examples of the use of epoxy adhesives could be quoted and these would include bonding facing panels such as stainless-steel sheets on to pillars and walls, bonding decorative aggregates on to panels and other substrates, joining extruded aluminium window frames together or the segments of an aluminium alloy foot-bridge. However, it is believed that sufficient examples have been given to enable the reader to appreciate the wide versatility of epoxy adhesives and the large savings that can often be made by using these materials and the new techniques associated with them rather than continuing with older well established methods. In other instances there has been no alternative to the use of an adhesive other than complete rebuilding of a structure.

REFERENCES

1. SHINGLETON, D. A., *Lecture to Adhesives in Building Conference, London* (May, 1969).
2. INTERNATIONAL UNION OF TESTING AND RESEARCH LABORATORIES FOR MATERIALS AND STRUCTURES (RILEM), *International Symposium, Paris* (1967).
3. *Epoxies with Concrete,* Publication SP-21, American Concrete Institute, Detriot (1968)
4. GAUL, R. W., and ZEIER, F. J., *Maintenance,* p. 35 (November, 1969)
5. WAKEMAN, C. M., STOVER, H. E., and BLYE, E. N., *Bull. Amer. Soc. Test. Mater.* **2**, No. 2, 93 (1962)
6. PARACEY, E., *Wld Constr.* **24**, No. 3, 32 (1971)
7. SHIMP, D. A., in P. F. BRUIN (Editor), *Epoxy Resin Technology,* Interscience, New York, Ch 8 (1968)
8. GAUL, R. W. and APTON, A. J., *Civ. Engng, N.Y.,* **29**, No. 11, 50 (1959)
9. SHELL INTERNATIONAL CHEMICAL COMPANY, *Epikote Age,* No. 9
10. *Plastics Institute Conference,* (June 1965)
11. HIRSCHI, T., *VDI-Ber.,* No. 122, 57 (1968)
12. ROBERTS, J. A., and VIVIAN, H. E., *Aust. J. appl. Sci.,* **12**, 104, 131, 428 (1961); **13**, 46 (1962)
13. TOPALOFF, B., *VDI-Ber.,* No. 122, 43 (1968)
14. *Araldite in Segmental Concrete Construction,* CIBA-Geigy
15. *Engineer, Lond.,* **221**, No. 5744, 325 (1966)
16. GRATTESAT, G., *Travaux,* No. 372, 1(1966); CHAUDESAIGUES, J., *Travaux,* No. 372, 5 (1966); MATHIVAT, J., *Travaux,* No. 372, 8 (1966)
17. DUNSTER, J. A. and HOUGH, N. E., *Civ. Engng publ. Wks Rev.,* **63**, No. 743, 651 (1968)
18. *Survr minic. City Engr,* **82**, No. 3981, 35 (1968)
19. SHELL INTERNATIONAL CHEMICAL COMPANY, *Resin,* No. 2
20. PINGUET, J. CL, and HOFER, R., *Route,* **53**, No. 7, 348 (1967)
21. *Construction des Viaducts de Chillon,* Bureau de Construction des Autoroute, Vaudoise, Bureau Technique Pinguet, Lausanne, and Consortium des Viaducs de Chillon, Lausanne, 1969
22. *Mater. Struct. Test. Res,* **1**, No. 3 (1968)
23. *Araldite in the Sydney Opera House,* CIBA-Geigy
24. *Araldite in a Sports Stadium,* CIBA-Geigy
25. KREIGH, J. D., NORBODY, G. M., and ENDERBROCK, E. G., *Highway Res. Rec.* No. 155, 1 (1967)
26. JOHNSON, R. P., *Struct. Engr,* **48**, No. 6, 227 (1970)
27. GAUL, R. W. and APTON, A. J. *Adhes. Age,* **3**, No. 5, 24 (1960)

14
Grouts for Levelling; Miscellaneous Applications

The precise and accurate location of heavy equipment and machinery during installation on concrete can be a very difficult operation, as some shrinkage, which is often uneven during the hardening of the concrete, is inevitable. But the use of epoxy-based grouting media has greatly simplified the operation and led to considerable cost savings. Hitherto, the spaces that remained between foundations and the under-surfaces of machinery were filled with a cementitious grout, but this procedure has a number of disadvantages. The grout:

(1) is relatively low in strength;
(2) has a high and unpredictable shrinkage;
(3) deteriorates in the presence of chemicals and oil; and
(4) is relatively slow to cure.

For these reasons, epoxy-based grouting compounds are now used throughout the world and they offer the following advantages:

(1) Very low shrinkage.
(2) Excellent adhesion to concrete and metal.
(3) Adequate strength in the grout is developed.
(4) Very good dimensional stability and mechanical strength and vibration damping. The epoxy grouts can be said, broadly, to have at least three times the compressive strength and four times the tensile strength of concrete.
(5) Good chemical resistance.

In practice, the preferred method is to cast the basic concrete structure and to leave a small gap of about ½ in (13 mm) between the concrete and the machinery that is to be precisely located. This gap is then filled with a free-flowing epoxy grout, the machinery having been positioned accurately before the epoxy was cured. Alternatively, the epoxy can be used to produce the desired accurately levelled surface on the concrete first and the machinery is placed in position afterwards. This technique has been used in the siting of heavy compressors and other machinery. Other examples are the installation of heavy milling, boring and drilling machines, which are used, for example, in the manufacture of marine diesel engines. These very heavy machines are normally mounted on a massive reinforced concrete foundation that is several feet thick. In certain cases, the machine is placed directly on to the prepared concrete base, accurately positioned and levelled, and the epoxy grout poured into a mould formed from, for example, foamed plastic strips around the base.

With very heavy machines, it has been necessary to support the levelled equipment temporarily with monolevelling jacks, form a mould as before, and pour the epoxy grout underneath each supporting jack. A similar method is used to support the rails on which the boring machine travels[1].

The method can also be employed in the regrouting of machinery, but the most impressive uses of self-levelling epoxy grouts have been concerned with tracks carrying heavy machinery. Two examples vividly illustrate this use.

The Harland and Woolf Shipyard in Northern Ireland has a dry-dock 150 m wide and 750 m long, which was built for the assembly of oil tankers from prefabricated sections, each of which weighs approximately 800 tons. These sections are carried from the workshops to their assembly position in the dry-dock by a giant crane 200 ft (60 m) high mounted on 60 bogies on each side and running on steel rails that extend down the length of the dock. The high dynamic mechanical stresses which occur when the crane is working would have caused the breakup of the concrete if the rails were put directly on to a concrete base. After considerable testing, it was decided instead to use a filled epoxy resin bedding composition, which possessed good tensile and compressive properties. The rails were carried on a steel sole plate 20 in (50 cm) wide, which was laid in 60-ft (18-m) sections. The bottom face of the plates was shot-blasted and primed with an epoxy primer under factory conditions. On site, the sole plates were aligned and levelled in place so as to allow a gap of 1–1½ in (2.5–3.75 cm) between the base of the plate and the concrete foundation. Shuttering was placed so as to allow a gap of 1–1½ in on either side of the plate and the resin grout was then deaerated and pumped into place so that the grout was level with the top face of the sole plate. About 2000 ft^3 (60 m^3) of epoxy grout were used in this one application[2].

The Jodrell Bank radio-telescope is used extensively for research in radioastronomy and also for tracking space vehicles. In this work, it is essential that the telescope is positioned accurately and maintained steady when aligned on its target. Additions and modifications to the moving parts of the installation have imposed heavier loads on the circular rail tracks on which the telescope rotates and it was found that the existing concrete under the sole plates was not strong enough to take these additional loads[3]. It was therefore necessary to fit new sole plates and to grout between these plates and the foundation concrete with a self-levelling epoxy grout, formulated to have high compressive and tensile strength. The telescope was fully operational within 7 days of placing the final section of the grout and the increased load is being supported satisfactorily. The work was completed without having to dismantle the telescope or its running gear and with a minimum disruption of the station's programme.

MISCELLANEOUS APPLICATIONS

In addition to the main uses already discussed, there are a number of other important applications for epoxy resins in the construction field, which are discussed below.

Soil consolidation

Oil wells drilled in loose sand can become blocked if the sand enters the well pipe, thus reducing the output of oil. To overcome this problem, a solvent-containing epoxy composition has been developed which can consolidate the sand in the immediate vicinity of the well bore. The consolidated sand then becomes a filter which prevents further movement of sand into the well itself. The technique of using this system first requires the water in a small area around the bottom of the well to be removed by means of an alcoholic solvent. The liquid resin system is then pumped into the pore space of the loose sand. Cure proceeds and the polymer spreads over the surfaces of the sand grains, concentrating at contact points between the grains. When the polymer is fully cured, the grains are firmly held together, forming a solid but permeable structure[4]. Using this technique, oil fields can be developed that were previously regarded as unworkable.

In a similar application, part of a Ruhr coalmine shaft has been sealed with an epoxy composition in order to prevent the ingress of water containing a high proportion of mineral salts[5]. Epoxy grouts were also injected into the rock walls of the underground NORAD Command Centre at Colorado Springs, U.S.A., in order to prevent rock movement[6].

Tile grouts

A point of weakness in ceramic tiling that is susceptible to chemical attack is the joint between the tiles. Normal grouts are often porous, have poor chemical resistance and adhesion and can allow bacteria to be harboured in their cracks. Epoxy tile grouts for walls and floors ensure that the chemical resistance of the tile is matched by that of the joints and also offer very much better adhesion and the absence of porosity and shrinkage. The Sidney Opera House contains about 80 miles (130 km) of ceramic roof tiles jointed with an epoxy system which completely excludes moisture and water vapour[7].

Epoxy laminates for concrete moulds

Wooden moulds and shuttering have been used for many years in casting and decorating concrete. However, there are a number of drawbacks to this use of timber:

(1) Much work is required to form a mould of complicated shape, which may only be used a few times before it becomes disfigured at the surface and be beyond repair. The construction of these moulds in timber also requires skilled manpower, which is expensive and difficult to obtain.

(2) Wood absorbs moisture from the wet concrete and there is therefore the risk of porosity developing in the concrete surface and the mould becoming warped.

Steel has also been used to make shuttering for concrete, but suffers from the disadvantages that it is very heavy, difficult and expensive to fabricate into complicated shapes and can easily corrode if it is not carefully maintained. Epoxy resin-based glass-fibre laminates are ideal materials for shuttering. The laminate is actually used as a surface or lining to timber moulds and shuttering. These epoxy-lined moulds are light, strong and extremely hard wearing and can be used many times. Smooth, flawless castings are obtained which reproduce perfectly the contours of the moulds.

Resin concrete

The term 'resin concrete' is used to describe materials in which resin, rather than Portland cement is used as the binder for aggregate particles. The resin serves the same function as the cement, initially providing a fluid matrix around the aggregate particles so that the mix can be compacted, and finally, when cured, it determines for the most part the properties of the material.

Various resins have been used to prepare resin concretes, including epoxies, polyesters, phenol–formaldehyde and furfurol–acetone types, and work on these resin mortars in Czechoslovakia has been described by Bares in a number of papers (reference 8 lists the earlier publications). Other studies have been carried out on epoxy resin concrete by Welch *et al*[9] and Lavie *et al*[10]. Towler and Williams[11] mentioned work on resin–sand mixtures in which they incorporated fibres of materials such as carbon, as a means of increasing the modulus of elasticity and improving the deformation characteristics of resin binders under sustained loads.

A further, different approach has been the addition of small amounts of resin into a normal cement concrete mix as a means of improving the tensile strength and other mechanical properties of the material.

Most of this work is still at the development stage and it is not at all clear whether a resin-based concrete will ever become a primary construction material. Nevertheless, the improvements shown by resin concretes over Portland cement concrete in terms of compressive and tensile strengths, elastic modulus, low porosity and chemical resistance will ensure that development work on the material will continue. However, resin concretes are expensive, long-term data on their mechanical behaviour are lacking and they are affected by increased temperatures and could not withstand a fire in the way in which cement concrete can.

The most likely applications for these mixes would appear to be as a speciality product for pipes, tanks, chemically resistant floors and perhaps as cladding panels for walls. Certainly in Czechoslovakia, large diameter pipes for the discharge of highly aggressive effluent from chemical plants have been made from a resin concrete, although the resin used was not epoxy but a furfural–furfurol type. In fact, at resin to aggregate ratios of 1:10 or greater, there is no clear case, on mechanical strength grounds, for using an epoxy resin rather than a polyester or a furfural resin in the concrete.

REFERENCES

1. *Araldite for the Grouting of Machinery,* CIBA-Geigy
2. *Epikote Grouting and Levelling Compounds,* Shell International Chemical Company
3. *Wimpey News,* No. 349, 4 (1971)
4. HAVENAAR, I. and MEIJS, F. H., *J. Inst. Petrol.,* **49**, No. 480, 382 (1963)
5. AU, E., *Min. Mag., Lond.,* **110**, No. 3, 171 (1964)
6. ERICKSON, H. B., *J. Soil Mech. Foundn. Divn., Amer. Soc. Civ. Engng.,* **94**, SMI, 159 (1968)
7. *Epikote Tile Grouts,* Shell International Chemical Company
8. BARES, R., *Chemy. Ind.,* No. 15, 482 (1970)
9. WELCH, G. B., CARMICHAEL, A. J. and HATTERSLEY, D. E., *Civ. Engng. pub. Wks. Rev.,* **57**, 671/672 (1962)
10. LAVIE, M. H., LERCHENTAL, H. and SPIRA, E., *RILEM Symp.,* Paris (1967)
11. TOWLER, R. J. and WILLIAMS, R. I. T., *CONRAD,* **1**, No. 4, 107 (1972)

SECTION V
Other Principal Applications

15

Composites

This chapter is concerned with the composite material produced when epoxy resins, acting as the continuous matrix, are reinforced with a fibrous material; in other words, epoxy laminates in which the reinforcement can be one of a number of materials. To most designers, the term 'reinforced plastics' usually means glass-reinforced plastics, often abbreviated to GRP, in which the resinous matrix is a polyester and the reinforcement is glass-fibre in one form or another. These laminates (or composites; in this chapter the terms are interchangeable) are widely used throughout the world for a multitude of purposes, such as translucent building panels, automobile bodies and boat hulls of various shapes and sizes.

Epoxy resin-based composites, on the other hand, have not been used to the same extent. For many applications, the properties provided by the polyesters are adequate and there has therefore been no desire to use instead the more expensive epoxy resins as matrix materials. However, there are some applications in which the superior properties of the epoxy resins are needed and in which the extra costs incurred by their use are offset by the benefits obtained. These applications are chiefly in the electrical industry for insulating tubing and printed circuits, in the chemical, oil and other processing industries for tanks and pipes, and in the aerospace industry for many parts of aircraft, rockets and space vehicles. It is also in this last industry where epoxy resins are at present the chosen system for use with the newer high-performance reinforcements such as carbon and boron fibres and 'whisker' materials of various types.

The basic principle underlying the development of the technology of composites is that the combination of two dissimilar materials results in a product with properties different from those of the starting materials. For example, glass-fibres can have a Youngs' modulus of about 10×10^6 p.s.i. (69 GN/m^2), an ultimate tensile strength of about 200 000 p.s.i. (1400 MN/m^2) and a linear stress/strain relationship to failure with no yield. Unreinforced epoxy resins, in comparison, have ultimate tensile strengths of 5000–15 000 p.s.i. (35–105 MN/m^2), Young's moduli of 0.2–0.5 x 10^6 p.s.i. (1.4–3.5 GN/m^2) and a non-linear stress/strain relationship. Combining the two materials into a composite results in a material with an ultimate tensile strength of about 60 000 p.s.i. (420 MN/m^2) and a Young's modulus of 4.0 x 10^6 p.s.i. (28 GN/m^2).

The properties of a composite depend upon a number of factors such as:

(1) The nature of the separate constituents and the adhesion between them. The fibrous reinforcement carries the major part of the applied stress in a composite, whereas the resin matrix transmits the stress to each fibre. The adhesion between the resin and the fibre is therefore the critical point, and if this adhesion is poor, rapid failure of the composite may occur when it is stressed.

(2) The relative amounts of resin and fibre present in the composite. In general, an increase in resin content leads to a decrease in strength, but an increase in resistance to weathering and chemical attack. A decrease in resin content imparts higher flexural and tensile strengths and Young's moduli. The electrical properties of epoxy–glass composites are hardly affected by the glass content except for the permittivity, which increases with increasing fibre content. Fibre contents depend upon the method used to fabricate the composite and, in general, filament-wound and compression-moulded 'pre-preg' materials have a fibre volume fraction of between 0.5 and 0.8, whereas hand lay up items have a lower fibre volume fraction (about 0.4).

(3) The orientation of the reinforcing fibres. They can be oriented in a unidirectional, bidirectional or random manner, and the properties of the composite will depend upon this orientation. Reinforced plastics do not show the same properties in all directions, i.e. they are anisotropic.

An example of this was quoted by Cornsweet[1], where the longitudinal tensile strength (i.e. the strength in the direction of the fibres) of a unidirectional composite consisting of equal volumes of boron fibre and epoxy matrix was 147 kg/mm^2 (1440 MN/m^2), whereas the transverse tensile strength was only 10.5 kg/mm^2

(103 MN/m^2). In a similar way, the tensile modulus decreased from 22 400 kg/mm^2 (220 GN/m^2) to 2100 kg/mm^2 (20.6 GN/m^2) when going from the longitudinal to the transverse direction. This highlights the need for designers to select combinations of fibre orientation that will provide the required degrees of strength in the various directions of the applied stresses in the design.

EPOXY AND OTHER MATRIX RESINS

Matrix materials for composites are not restricted to synthetic organic polymers. Successful composites have been fabricated from inorganic matrix materials such as phosphate-bonded oxides, cermets (ceramic–metal combinations) and metals themselves. These composites are not within the scope of this chapter, which is restricted to the consideration of thermosetting organic resins as matrices.

The factors that usually need to be taken into consideration when choosing a matrix resin are:

(1) The mechanical strength required from the composite. This strength also depends upon the type of reinforcement and its arrangement within the matrix.
(2) The specified performance of the composite in its working environment. The matrix must be able to maintain its properties under service conditions.
(3) The fabrication process to be used. The resin should be easy to use in the process.
(4) The cost of the resulting composite in conjunction with the use to which it will be put.

Five resins are currently used for the production of composites, other than for research and development purposes:

Epoxies
Polyesters
Phenolics
Silicones
Polyimides

Of these five, polyesters are the most widely used, followed by phenolics. Epoxy resins probably rank third in terms of amounts used, followed by silicones and then polyimides. Although lower

Table 15.1

ASSESSMENT OF MATRIX RESINS (AFTER LIN[3])

Resin	*Mechanical properties*	*Processability*	*Long-term heat resistance*	*Short-term heat resistance*	*Electrical properties*	*Chemical resistance*	*Cost*
Polyester	Good	Very good	Poor	Fair	Very good	Good, especially to acids	Low
Epoxy	Very good	Very good	Poor	Fair	Good	Good, especially to alkalis	Low
Phenolics	Very good	Good	Fair	Good	Fair or good	Good, especially to acids	Low
Silicones	Fair or poor	Fair	Good	Very good	Very good	Good	Moderate
Polyimides	Good	Fair or good	Good	Very good	Fair or good	Good, especially to acids	High

in cost than epoxy resins, polyesters are much inferior in mechanical properties. Silicones are expensive low-strength resins and are chosen only when their long-term heat resistance and electrical properties are needed. Phenolics offer good short-term performance at high temperature and are low in price. The more recently developed polyimides are used almost exclusively for their short- and long-term thermal resistance properties. Generally, epoxy matrices are used for continuous exposure to temperatures of 175 °C, and polyimides for 175–310 °C. For short-term exposure, epoxy resins are acceptable up to 230 °C, whereas polyimides can be used up to 370 °C[2], but imides are difficult to process as they require a high curing temperature and release volatile compounds during cure. A summarised assessment of these five resins by Lin[3] is given in *Table 15.1.* Although this table was drawn up in 1968, and was constructed from the point of view of the U.S. aerospace industry, the relativities it expresses can be regarded as still valid for current industrial applications of composite materials.

The properties of epoxy resins that lead them to be selected for high-performance applications of composites are:

(1) Easy handling and processability.
(2) Low shrinkage and absence of volatile compounds during cure.
(3) Good adhesion to glass and other reinforcing materials.
(4) High level of thermal resistance.
(5) Excellent chemical and moisture resistance.
(6) Good thermal and electrical insulation properties.

Partly resulting from these properties, epoxy composites in which glass, carbon or boron fibres are used as reinforcements have a very high level of mechanical properties and fatigue resistance and chemical and thermal resistance. Values for these properties are given later in the chapter.

FIBROUS REINFORCEMENTS

The technology of plastics reinforced with glass-fibre is well established, and epoxy GRPs have a wide variety of applications. Examples mentioned later include their use in rocket motor cases for the Minuteman, Polaris and Poseidon missiles, and for many secondary structural components of aircraft such as radomes and fairings. They have been used very little for primary load-bearing structures, partly because of their lack of stiffness. Epoxy–glass is superior to metals in strength/weight (specific strength) characteristics, but offers no advantages in stiffness/weight (specific Young's modulus). Therefore, although epoxy composites have an acceptable performance for applications such as filament-wound pipes, tanks and printed circuits, their use in aerospace and as engineering materials alongside metals, especially for primary load-bearing structures, has been impeded by the lack of a strong and stiff fibre reinforcement.

This need led directly to the development of new stiff fibres such as those of carbon and boron, and continues to encourage research into more exotic fibres and 'whiskers'. Whiskers have been defined[4] as single crystals of materials such as aluminium oxide that have a length to diameter ratio of at least 10:1 and a maximum cross-sectional area of $7–9 \times 10^{-5}$ in^2. These very short, discontinuous fibres can be used as a reinforcement in their own right or to improve the properties of continuous fibre composites.

Table 15.2

STRENGTHS OF VARIOUS FIBRES[6]. (Courtesy of the Council of the Institution of Mechanical Engineers)

Fibre	*Ult. tensile strength*		*Young's modulus*		*Specific gravity,*	*Specific modulus,*
	p.s.i. × 10^3	MN/m^2	p.s.i. × 10^6	GN/m^2	g/cm^3	p.s.i. × 10^6
Boron	360	2520	60	420	2.50	24
Carbon (RAE Farnborough, U.K.) (high modulus)	300	2100	60	420	2.00	30
E glass filament	250	1750	10	70	2.54	3.9
High tensile steel	190	1330	30	210	7.87	3.8
Aluminium alloy	67	469	10.5	73	2.80	3.7
Titanium alloy	134	938	16	112	4.50	3.5

Carbon and boron fibres are at present the most widely used reinforcements for advanced composites. The strengths of these fibres are shown in *Table 15.2* in comparison with those of some competing metals[5,6], and the data indicate the superior specific stiffness possessed by fibres of carbon and boron compared with the metals. When these fibres are incorporated into a composite with an epoxy resin, the order of strengths of the materials remains roughly the same (see *Table 15.3*). Another way of depicting these relative strengths is to plot the specific tensile strength (tensile strength/density) as a function of the specific tensile modulus. This has been carried out in *Figure 15.1*. The more ideal are the mechanical properties of the material the more closely it will tend towards the upper right-hand corner of the graph. This is the area where high strength and stiffness are combined with low density.

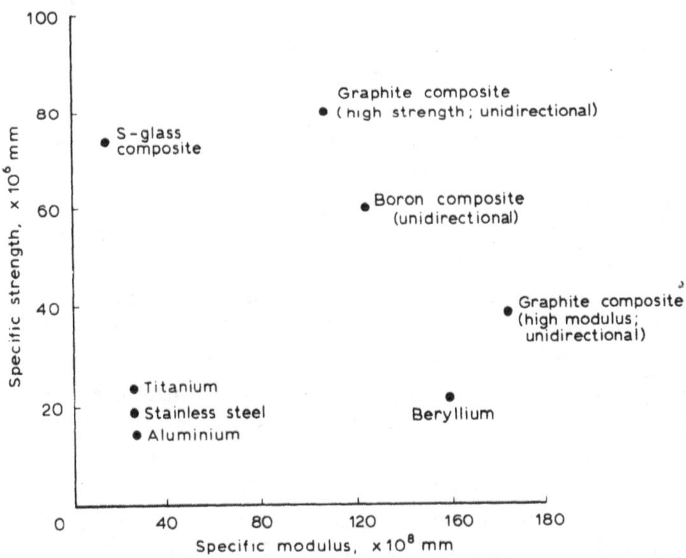

Figure 15.1. Comparative strength properties of various reinforcements[1]

It is apparent from *Figure 15.1* that although an epoxy–glass composite is clearly stronger than any conventional metallic material, it is not stiffer. Similarly, beryllium is much stiffer than the conventional metals but no stronger in specific tensile strength. The carbon- and boron-based unidirectional composites are clearly superior both to the metals and to the glass-based composite in strength and stiffness per unit weight.

DESIGNING WITH FIBRE-REINFORCED EPOXY RESINS

Designers who wish to use epoxy composites, or indeed composites based on other resin matrices, should be fully aware that they are dealing with an anisotropic material which has to be designed and constructed itself, as well as the structure in which it is to be used. The properties of the material need to be kept firmly in mind and a direct replacement copy of an already existing component made in metal should not necessarily be attempted. Because the designer is able to determine the properties of the composite itself, he has an additional degree of freedom over working with metals. He also has the additional burden of obtaining an understanding of the technologies concerned with combining fibre with matrix, of designing the component to take all the likely imposed stresses, and of fabricating the component by methods not normally familiar to the engineering industry.

The cost of using an epoxy composite rather than a more traditional material is of great importance, and must be set against the likely benefits and disadvantages that will accrue. A direct comparison of raw material costs without reference to any property is misleading. Similarly, any weight savings should not be considered as a sole justification for using the composite, if the other opportunities for improvements and benefits are not considered. With this problem in mind, Robbins[7] introduced the concept of 'cost per unit of property per unit of length' as a useful comparison figure for materials. This value, he claims, gives the specific cost of strength or stiffness and provides a true comparison of the cost of achieving a given value of that property for different materials. On this basis, the epoxy–carbon composites are an order of magnitude more expensive than the conventional engineering materials *(Table 15.3). Figure 15.2* illustrates these data in the form of a block diagram.

Composites of epoxy resins with carbon or boron fibres possess properties that are not commonly found in engineering materials. They have high strength, low density, high Young's modulus and good elasticity, showing only ½–1 per cent elongation at break. Steel, in contrast, shows a 20 per cent elongation at failure. A carbon-reinforced composite can be loaded to 75 per cent of its ultimate tensile strength, and will recover when the load is removed; a metal would be permanently deformed at this loading level. As far as mechanical properties are concerned, therefore, this elasticity factor

Table 15.3

COMPARATIVE COSTS AND PROPERTIES (FROM ROBBINS[7])

Material	*Ultimate tensile strength,* MN/m²	*Young's modulus,* GN/m²	*Specific gravity, g/ml*	*Selling price,* p/kg	*Cost/ unit strength/ unit length,* p/MN/cm	*Cost/ unit stiffness/ unit length,* p/GN/cm
Steel En19	999	207	7.8	13.77	1.07	5.17
Mild steel	617	207	7.8	9.83	1.23	3.67
Aluminium alloy HV 30 WP	345	69	2.7	77.19	6.03	30.10
Die-cast aluminium	247	72	2.7	29.50	3.24	11.11
Magnesium alloy Z.W.1.	276	44	1.8	212.41	13.84	86.81
Die-cast zinc	277	74	6.6	24.58	5.78	21.62
Titanium DTD 5173	931	110	4.5	590.03	28.46	24.09
Glass-filled epoxy	494	–	1.6	405.15	13.16	–
Glass-filled nylon	154	8	1.3	98.34	8.44	108.33
Glass-filled polyester	386	3	2.1	309.76	16.84	2166.67
Acetal	69	4	1.43	76.70	15.94	275.00
Polycarbonate	55	2	1.2	76.70	16.76	450.00
Nylon 66	62	1	1.13	58.61	11.29	700.00
C.F.R.P. epoxy^{HT}	1103	117	1.6	9178.19	133.18	1255.56
C.F.R.P. epoxy^{HM}	813	207	1.5	9178.19	169.37	665.22
C.F.R.P. epoxy^{A}	–	–	1.5	7342.51	–	–

enables advanced composites to be loaded to a much higher proportion of their ultimate strength than orthodox alloys and metals.

In general, designers should capitalise on the high strength that composites possess along the lines of the fibre axes, although in articles where the stresses are in more than one direction this can be very difficult to achieve. The design problem in articles where there is a definable stress pattern in one direction is relatively simple and the anisotropic nature of the composite allows the maximum strength properties to be fully utilised.

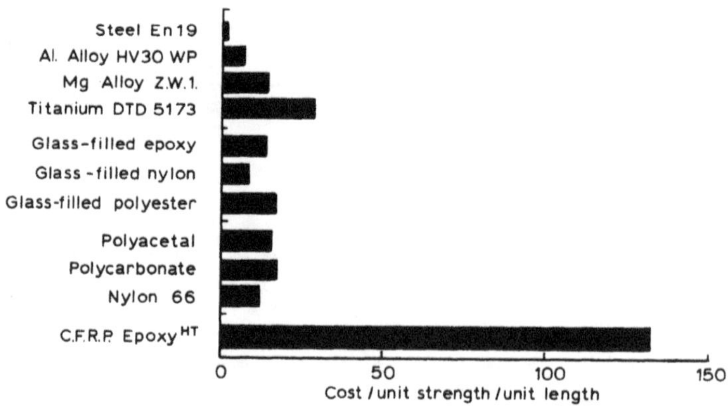

Figure 15.2. The specific cost of strength in commonly used materials (From Robbins[7])

An important factor possessed by composites is their ease of fabrication into complex shapes. This can lead to significant cost savings over the 'all-in' cost of items made from traditional materials. Limiting factors to the wider use of epoxy composites are the upper temperature limit to their use compared with those of most metals, and the general lack of data on their performance suitable for design work. Single point data for mechanical properties which might serve to characterise a metal are not sufficient to characterise a reinforced polymer. The need is for data that cover, for example, a range of strains and stresses made at different rates and for different periods of time and taking into account any relevant environmental effects.

SOME APPLICATIONS OF EPOXY-BASED COMPOSITES

Epoxy resin-based composites are used where their superior properties are needed and the extra cost involved in using them is offset by the benefits obtained. A brief account of some of their chief applications in different industries is given below.

Oil industry

The transport, storage and processing of oil and oil products requires tanks and pipes that can withstand the corrosive attack of these substances. A similar requirement exists in the chemical, food, gas and water industries, all of which have a need for chemically resistant tanks and pipes. Made by the filament winding technique, there are now epoxy–glass tanks throughout the world being used to store acids, alkalis, chlorinated brine, salts, oils, etc. These epoxy tanks can be produced at a lower initial cost than other comparable tanks except those made from unprotected carbon steel. Moreover, the wound epoxy tanks require less maintenance, are lighter in weight and are more corrosion resistant than steel tanks. A further development is the fabrication of filament-wound tanks, such as large-capacity vats and acid storage containers on-site[8]. Prefabricated inner shells may be used as the moulds, which can be collapsed during transit between locations. Vats 28 ft in diameter and 10 ft high have been built in this way.

Some data have been given by the U.S. tank manufacturers Black, Sivalls and Bryson, and were quoted by Reeves[9] in 1963, of cost comparisons for filament-wound tanks of 12 000 gal capacity, made from different materials. These data are shown in diagrammatic form in *Figure 15.3* and demonstrate the cost advantage that epoxy tanks commanded in the U.S.A. at that time. However, nothing has happened in the last decade to cause the advantage to revert to metals. On the contrary, epoxy resin prices have softened considerably whilst steel prices have hardened.

Road and railway tank cars are also a special example of the use of epoxy–glass filament-wound structures[5]. One rail tank car built in this way was 55 ft long, had a diameter of 9 ft, a weight with carriage of 26 tons and a carrying capacity of 22 500 gal. The weight of the epoxy tank alone was 8000 lb and replaced a steel tank that weighed 27 000 lb, a saving of 19 000 lb. The inner lining of the epoxy tank was an epoxy–asbestos laminate for maximum chemical resistance. The initial cost of the epoxy tank was less than those of an aluminium, stainless steel or lined steel tank, and the weight saving achieved enabled a tank with a greater capacity to be fitted

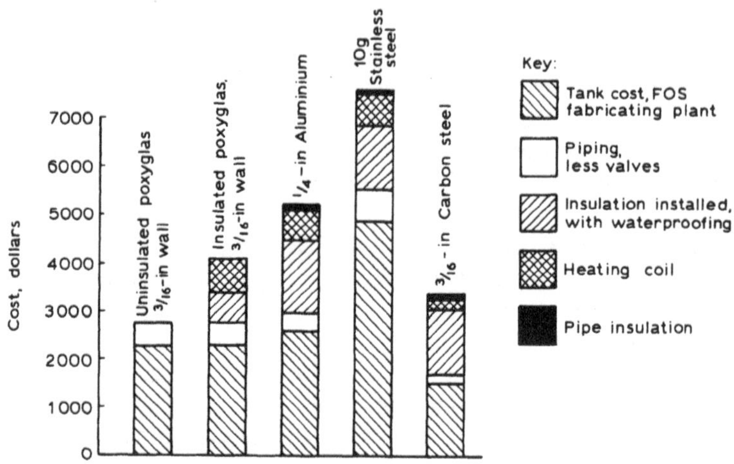

Figure 15.3. Cost comparison for 12 000 gal storage tanks made from different materials (From Reeves[9])

into the original truck. The epoxy tank car has therefore a higher payload to weight ratio, and also offers outstanding corrosion resistance and minimum maintenance at a competitive price to steel, plus easy cleaning and good thermal insulation properties. Naturally, to avoid possible large electrostatic charges developing during loading and discharging, the glass–epoxy tanks should be fitted with internal collector plates which can be properly earthed.

Extensive use of epoxy–glass filament-wound pipe is now made in the U.S.A. and Europe not only for services such as water, natural gas and petroleum, but also for gas gathering, brine lines and waste disposal. In general, reinforced plastic pipes have four attractions:

(1) They are considerably lighter than any metal pipe of comparable diameter and rating, and can be designed for a specific rating by orientation of the glass-fibre.
(2) They can withstand corrosive attack from most chemicals, and require minimum maintenance.

(3) Their smooth internal surface reduces friction loss and the build-up of solid materials on the pipe walls.

(4) They can operate at a temperature of 120 °C.

Considering the U.S. position, the raw material costs for epoxy pipe are lower than those for metal. The weight of a resin pipe is also less than that of a comparable length of metal pipe, and this is reflected in lower transport and installation costs for the epoxy pipe.

Epoxy resins have a serious competitor in polyesters as far as piping is concerned, and one forecast of the future size of the U.S. market for reinforced plastic pipe gives the two types of resin about equal shares of the market. Polyesters are no doubt favoured for their acid resistance and cheapness, whereas the short- and long-term strength of epoxy resins are clearly needed, coupled with their chemical resistance, for use in the difficult situations that occur in oil fields. These data are given in *Table 15.4* and it can be seen that the oil industry is forecast to be the major user of epoxy pipe, followed by chemical processing and gas distribution, and other, miscellaneous users. This forecast growth in the consumption of epoxy pipe will, however, depend upon the price of the pipe remaining competitive with steel pipes. High-speed filament winding techniques now operating and the development of vinyl–epoxy resin combinations will all serve to keep prices down and to ensure the future demise of the metal pipe.

It should be noted however, that at present (1973), because of their limited resistance to impact and eventual disintegration in a fire, glass-reinforced epoxy pipelines are not considered to be suitable for above-ground service with flammable products. Moreover, high electrical charges with sufficient energy to ignite flammable mixtures may develop on the outer surface of a glass–epoxy pipe. Although steel lines are therefore the first choice for above-ground installations, in situations where lines must be buried, glass–epoxy pipes are normally suitable and safe, even for flammable products. The epoxy pipes also have the advantages of outstanding resistance to internal and external corrosion, but naturally precautions must also be taken to conduct to earth any charges that develop on the surface of the pipe when pumping products of low conductivity.

Above-ground storage tanks made from glass–epoxy combinations are also not recommended for flammable products, the objections

Table 15.4

U.S. MARKET FOR REINFORCED PLASTIC PIPE (AFTER WOOD[10])

Market	Millions of lb								
	1965			1970			1975		
	Polyester	*Epoxy*	*Other*	*Polyester*	*Epoxy*	*Other*	*Polyester*	*Epoxy*	*Other*
Chemical processing	3.2	1.7	0.4	6.4	4.0	1.1	23.0	12.0	5.0
Oil industry	N	0.1	N	N	4.5	2.0	3.0	30.0	5.0
Gas distribution	N	0.1	N	N	0.7	0.1	2.0	9.0	2.0
Water distribution	N	N	N	10.0	N	N	30.0	N	2.0
Mining industries	0.4	0.1	N	1.2	0.2	N	3.0	0.5	N
Miscellaneous, including fittings	0.7	1.2	0.1	1.7	2.7	0.4	9.0	12.5	2.0
Total	4.3	6.1	0.5	19.3	12.1	3.6	70.0	64.0	16.0

The data include pipes based on glass and other reinforcements. The 'Other' category includes vinyl epoxy, vinyl ester, phenolic and furan systems.

N = Negligible (< 100 000 lb).

to their use being similar to those for glass–epoxy pipes. Likewise, the epoxy tanks are eminently suitable for non-flammable products above-ground, and for both types of product below-ground, especially where the expected life of a steel tank, due to corrosion, is less than 15 years.

Electrical industry

In this industry, the most sought after properties of epoxy composites are mechanical strength and electrical resistance, and many examples are available to illustrate the successful use of these materials. Epoxy–glass slot liners are used in certain large electrical generators. The liner must have good mechanical and electrical properties and must be dimensionally accurate to fit the slots. Large air-blast circuit-breaker tubes (see Chapter 8) rely on epoxy–glass tubes for structural and high-pressure storage purposes within the interrupters and the main support insulators, and an epoxy resin composite was chosen because of its excellent mechanical and electrical properties. Filament-wound rings are used extensively for the support of stators in turboalternators, and low-tension fuse links for outdoor use may be enclosed in an epoxy–glass tube and the ends sealed with epoxy resin to give complete protection to the fuse from adverse environments. Epoxy–glass composites have also been widely used by the U.K. Science Research Council in the construction of vacuum vessels for the electron synchrotron NINA.

Aerospace industry

This industry was the first to exploit to the full the strength to weight advantages of epoxy and other composites and is one where the cost of the raw material is less important than the weight saving that the material might achieve. Space vehicles have many different components constructed in epoxy composite materials. For example, power source packages, structural spheres, antennae, and rocket motor bodies are almost all filament wound with an epoxy–glass system. Over 30 development and production missiles in the U.S.A. have been made in this way. Other parts of the rocket, such as the nozzles, exhaust cones, blast tubes and radomes are also constructed of epoxy–glass composite. Some U.S. Army data published in 1962[11] *(Figure 15.4)* show a cost comparison for the production of rocket motor cases from various materials, and the epoxy composites are

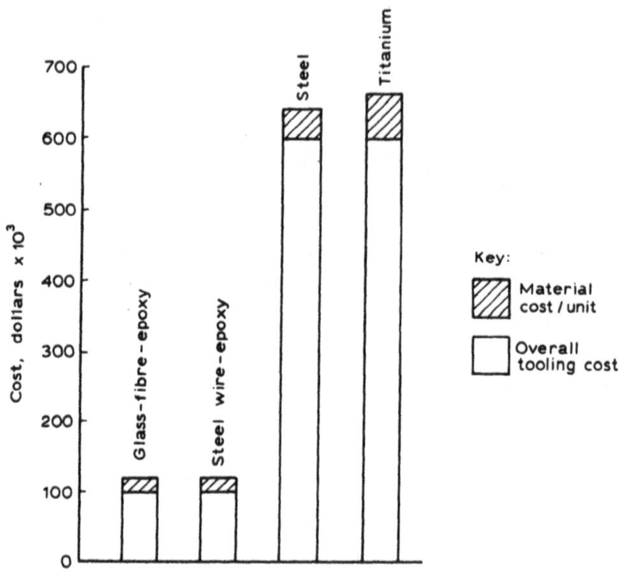

Figure 15.4. Cost comparison for rocket motor cases (From Shibley *et al.*[11])

clearly considerably cheaper than the steel or titanium cases. Several U.S. missiles have had their performance improved by the substitution of an epoxy composite for a metal component. Savings of 90 per cent of the cost of the cases for Minuteman were made by replacing titanium in this way. Similarly, the range of Polaris was increased, and weight and cost savings of 25 per cent were achieved for the Atlas missile.

Aircraft industry

Together with the aerospace industry, the aircraft industry has shown most activity in using composites on a production basis. Again, as with space vehicles and missiles, this is due to the weight savings being readily translated into cash savings, which offsets the high cost of the raw materials. There are very many examples of the successful use of epoxy composites in aircraft, but it is possible to consider only a small number of these here.

Epoxy–glass aircraft propeller spinners are used because of their high strength to weight ratio and high resistance to accidental damage and corrosion. Anti-icing pads for the intake of Rolls-Royce Tyne and Dart engines are made by embedding heating elements into epoxy–glass laminates. The material is very resistant to abrasion and to the required temperature changes that take place.

However, the most exciting work in composite technology today is taking place in the aircraft industry with the development of the so-called 'advanced composites'. These composites are based at the moment on fibres of graphite (carbon) and boron. The properties and advantages offered by composites based on these fibres have been mentioned already, and some examples of their use are given below.

Aircraft flooring is an area where carbon fibre-reinforced epoxy (CFRP) systems have a good opportunity to replace the aluminium honeycomb floor panels that are currently used[12,13]. Rolls-Royce (Composite Materials) Ltd. (RRCM) has carried out considerable development work with BOAC on a floor panel consisting of a CFRP skin bonded to a nylon paper-based honeycomb (Nomex) core. These panels have been fitted into two BOAC VC-10s, two BOAC Boeing 747s, a BEA Trident, two Lufthansa Condor 747s and two United Airways Boeing 747s. Taking into account factors such as the cost of materials, the life of the panel and the cash value of weight for a particular aircraft, after 2 years' service the costs of the CFRP–Nomex panels (at 0.48 lb/ft^2) appear to be equal to aluminium balsa (at 0.82 lb/ft^2) panels. At 5 years' service, there is a saving of just over £2/ft^2 of floor panelling per year, which, for a Boeing 747, is £6000 per year over 5 years. The life of the CFRP panel has been assessed to be equal to that of the aluminium panel at half the weight[14]. Further, if all gangways and underseat panels in a Boeing 747 were replaced with CFRP panels, an extra seven passengers could be carried.

Other CFRP uses in aircraft have been the fabrication of the complete wing spars of the T59C Kestrel Slingsby Sailplane. This I-beam, which runs the length of the wings and supports the whole weight of the aircraft, has enabled the wing span to be increased. Many other examples exist in powered aircraft where CFRP can make substantial weight savings, and several of these were mentioned

Table 15.5

ADVANCED COMPOSITE FLIGHT TEST PROGRAMMES (FROM JUNE AND KELLY[15])

System	*Component*	*Composite Material Concept*	*Remarks*
General Dynamics F-111	Horizontal stabiliser	Boron–epoxy skins on full depth honeycomb core	27% weight saving from aluminium baseline
	Wing trailing edge panel	Boron–epoxy	22 installed on operational aircraft
McDonnell Aircraft F-4	Rudder	Boron–epoxy on full depth honeycomb core	45 installed on operational aircraft, 35% weight saving
Grumman F-14	Horizontal stabiliser	Boron–epoxy skins on full depth honeycomb core	Baseline design. First advanced composite component committed to production. Currently flying as part of F-14 flight test programme. 19% lighter than titanium alternative.
Lockheed C-5A	Wing leading edge slat	Boron–epoxy skins stabilised by honeycomb; on supporting ribs	7 installed on operational aircraft. 21% weight saving. 5000 h total time, with over 1000 h on high time part
Sikorsky S-61	Tail rotor blades	Boron–epoxy spar, fibre-glass epoxy skins	First flight test of primary structure incorporating advanced composite. 5% weight saving
Douglas Aircraft A-4	Wing flap	Boron–epoxy skins on full depth core	32% weight saving. Part subsequently fabricated with graphite–epoxy with 47% weight saving
Boeing 707	Wing foreflap	Boron–epoxy skins, thin honeycomb monocoque	25% weight saving. FAA certified. 2 in commercial service with 5000 flight hours on each
Boeing 737	Spoiler	Graphite–epoxy skins on full depth honeycomb core	20% weight saving. FAA certified. 2 installed in commercial service in late 1971
Boeing CH47	Main rotor blades	Boron–epoxy	

by Thomas[14]; for example, an aileron control strut for the VC-10, an intermediate transmission shaft for the Westland Wasp helicopter and a flap-operating torque tube for the Hawker Siddeley HS-748. It is also hoped to develop a horizontal tail boom for the Wasp that will save 28 per cent of the weight of the existing metal boom. A Hovercraft propeller has also been moulded in (epoxy) CFRP, and the weight reduction that this achieved enabled a smaller hub and pitch change mechanism to be used.

In the U.S. aircraft industry, very many trials are now under way on the use of carbon- and boron-reinforced epoxy systems and some composites of this type are already being included in production line aircraft. Wing flaps for the A-4 aircraft have been developed in boron and carbon and boron–epoxy composites are being used or evaluated for many different components. *Table 15.5*, taken from a paper by June and Kelly[15], summarises the advanced composite test programme extent at the beginning of 1972.

Finally, it is interesting to look in a little more detail at the three components of the F-111 swing-wing aircraft that are fabricated from boron–epoxy composites[16]. These components are:

(1) Wing trailing edge panel.
(2) Wing pivot fitting doubler.
(3) Wing pivot fitting fairing.

The wing trailing edge panel provides an early (1967) example of the use of a boron–epoxy composite in a production line situation. The panel has a span of 70 in, chord of 25 in and thickness of 2 in and is wedge-shaped. The design is similar to the original aluminium panels which it replaced, but contains some parts in a glass–epoxy composite, which was also used to achieve a smooth surface transition between the boron–epoxy panel and the aluminium wing. Overall, the cost of the boron–epoxy modification was greater than the cost of the 100 per cent aluminium wing, but a 16 per cent weight saving over the aluminium design was achieved. A boron–epoxy composite was chosen for the wing pivot fitting doubler, in order to reduce the stress levels in the steel wing pivot fitting, which is a minimum weight, highly stressed structure. In this way, it was hoped to increase the fatigue life of the steel component. The boron–epoxy doubler is a 24-ply laminate bonded direct to the lower steel plate of the doubler, and has been found to reduce stresses in the critical areas of the steel structure and hence double the fatigue life.

The alternative to fitting the doubler would have been to redesign the entire steel wing pivot unit. This would have resulted in an intolerable delay and additional large tooling costs. The doubler allowed the modification to be made to new aircraft at a 21 per cent saving on the cost of redesigning in steel *(Figure 15.5)*. For existing aircraft, the saving in using a boron–epoxy doubler rather than fitting a redesigned steel unit is 60 per cent. The use of the

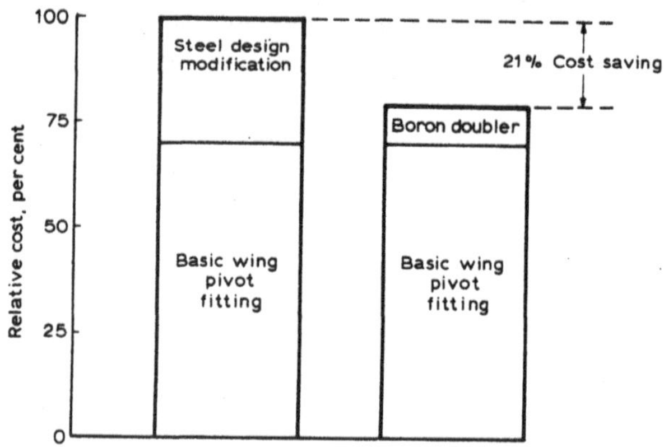

Figure 15.5. Relative costs of redesigning a steel wing pivot unit and fitting a boron–epoxy doubler[16]

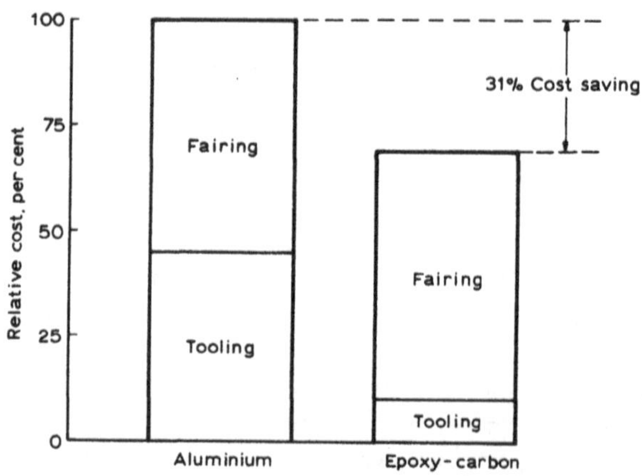

Figure 15.6. Relative costs of a fairing for a wing pivot fitting using aluminium and a carbon–epoxy composite[16]

boron–epoxy doubler required the redesign of the wing pivot fitting (WPF) fairing to make it detachable in order that regular inspection could be carried out. The original fairing was a glass-fibre sandwich bonded directly on to the WPF plate. The new detachable fairing was made from a 12-ply carbon–epoxy composite stiffened with a honeycomb core, the alternative being an aluminium-faced honeycomb sandwich panel. The relative costs of the two are given in *Figure 15.6,* where a 31 per cent saving in cost achieved by using the epoxy composite is apparent. The epoxy composite also offered a 26 per cent saving in weight over the aluminium panel.

Sports equipment

Selling in a market where price is not as important as product performance, sporting equipment manufacturers have been quick to realise the advantages that advanced composite materials can bring to their products. Skis were one of the first items of equipment to incorporate an epoxy–glass composite and studies are now well under way to incorporate CFRP in them.

Stronger and lighter squash rackets can be manufactured using CFRP veneers and a wood core. Designers have been able to reduce the weight of a golf club by 50 per cent while still retaining acceptable club performance, but using CFRP shafts. Racing oars have been made 30 per cent lighter, 25 per cent less in side-face area and 10 per cent stiffer than a standard wooden oar, by incorporating four CFRP reinforcing bars into a wooden core and wrapping a CFRP 'pre-preg' around the outside of the oar.

Small, carefully sited strips of CFRP 'pre-preg' have been found to be effective in increasing the stiffness of GRP bodies of kayaks and canoes. Alternatively, the existing stiffness can be retained and the amount of GRP reduced. In this way, weight savings of up to 25 per cent have been achieved. Cricket bat springs and yacht masts have also incorporated CFRP to improve stiffness and fatigue life.

Production engineering

One of the first epoxy–carbon composite applications outside the aircraft industry was for the bow of a wire-stranding machine[14]. The machine is used for twisting wires together for the manufacture of electrical cables, and the production rate is dependent on the speed of rotation of the bow. An increase from 500 to 2000 r.p.m. was achieved by constructing the bow from CFRP rather than steel.

An area where epoxy–carbon fibre composites have great potential is in the fabrication of small reciprocating machine parts such as shuttles and other parts of weaving machines, the revolving parts of centrifuges and parts of internal combustion engines. In these cases, it is possible to fabricate the moving part out of CFRP so that it has the same mechanical strength but weighs considerably less than if the part were constructed in metal. The force required to overcome inertia and move the part is therefore reduced, or faster movement can be obtained for the same energy input. In one example, a 50 per cent increase in speed of working of a needle loom was obtained by using a CFRP frame rather than the original aluminium frame.

Glass-reinforced epoxy systems are also widely used for checking figures and fixtures, and as moulds for GRP production. These uses are fully described in Chapter 18.

The future for epoxy composites

Inevitably, in the long term composites will become primary construction materials in the building and engineering fields and replace metals in very many applications. Already we are seeing this take place in piping, aircraft, missiles and satellites and there are signs that advanced composites are beginning to be used in the general engineering field alongside metals, and they will continue to be found wherever weight must be saved without loss of stiffness.

This trend will continue and accelerate if the economics continue to be favourable towards reinforced plastics, which will depend upon the continued improvement of fabrication techniques as well as changes in raw material prices. A further important factor is the education of designers in the use of composite materials which, as has been remarked earlier, are very different from orthodox materials in both properties and fabrication methods.

Technically, the efforts made towards developing other high-performance fibres and matrix resins will continue, but this will be sustained mainly by the needs of the aerospace industry. In general engineering and other industries, the emphasis will be on exploiting existing technology and unusual combinations of materials, for example combinations of foam and composites. An example which typifies the latter type of approach is the 'foam reservoir moulding' technique (FRM), recently developed by Shell[17]. This is a process which produces a composite structure using a thermosetting resin (typically an epoxy resin), fibrous reinforcement and a sheet of flexible polyurethane foam. The foam is impregnated with the resin system and laid in a prepared mould with the dry fibrous reinforcement placed between the foam and the mould surfaces, top and bottom. On closing the mould, the resin mixture is squeezed out of the foam and impregnates the reinforcement, the whole composite taking up the shape of the mould. The resin system cures and, on opening the mould, a rigid composite structure is obtained of the shape of the mould. There are many advantages of this technique. The process itself is simple and does not require expensive equipment, and the product has a high stiffness to weight ratio and can be designed to have a wide range of characteristics.

Epoxy resins can be used for this and all of the other techniques used at present for composite production. Whether an epoxy or another resin is chosen for the resinous matrix in any situation will depend upon a cost/performance assessment of the composite produced using the various resin combinations. For high-performance applications, typically those in the aerospace industry, it seems likely that epoxy resins will retain their position on performance grounds. At the lower level of performance, where epoxy resins and high performance polyesters overlap in many of their properties, the choice of resin is less clear, and has been made more difficult by the development of vinyl–epoxy resins. This situation will no doubt continue in the future.

REFERENCES

1. CORNSWEET, T. M., *Science, N.Y.*, **168**, No. 3930, 433 (1970)
2. FROST, J. D., HERTZ, J. and CHRISTIAN, J., *SAMPE Q.*, **3**, No. 2, 34 (1972)
3. LIN, H., *Plastics Polym.*, **36**, No. 123, 249 (1968)
4. *Design Engng.*, **18**, No. 4, 51 (1972)
5. ROSATO, D. V. and GROVE, C. S., *Filament Winding*, Interscience, New York (1964)

6. LANGLEY, M., *Chart. mech. Engr.*, **17**, No. 2, 56 (1970)
7. ROBBINS, D. K., *Annual Tech. Conference, SPI,* (1971) Section 13-A
8. *Chem. Engng,* **70**, No. 6, 210 (1963)
9. REEVES, J. F., *Annual Tech. Conference, SPI,* (1962) Paper 8C
10. WOOD, A. S., *Mod. Plast.*, **47**, No. 7, 74 (1970)
11. SHIBLEY, A. M., PERRITT, H. L. and EIG, M., *Plastec Rep., U.S. Army,* No. 10 (May, 1962)
12. *Design Engng,* **18**, No. 1, 85 (1972)
13. *Mod. Plast.*, **2**, No. 2, 40 (1972)
14. THOMAS, J. A. G., *SAMPE Q.* , **3**, No. 2, 45 (1972)
15. JUNE, R. R. and KELLY, J. B., *SAMPE Q.*, **3**, No. 2, 58 (1972)
16. DIAL, D. D. and HOWETH, M. S., *SAMPE Q.*, 3, No. 1, 17 (1972)
17. *Foam Reservoir Moulding Technique,* Shell International Chemical Company

PLATE 10
An epoxy mortar was used to butt joint sections of the Chillon Viaduct, Switzerland (Courtesy of CIBA-Geigy)

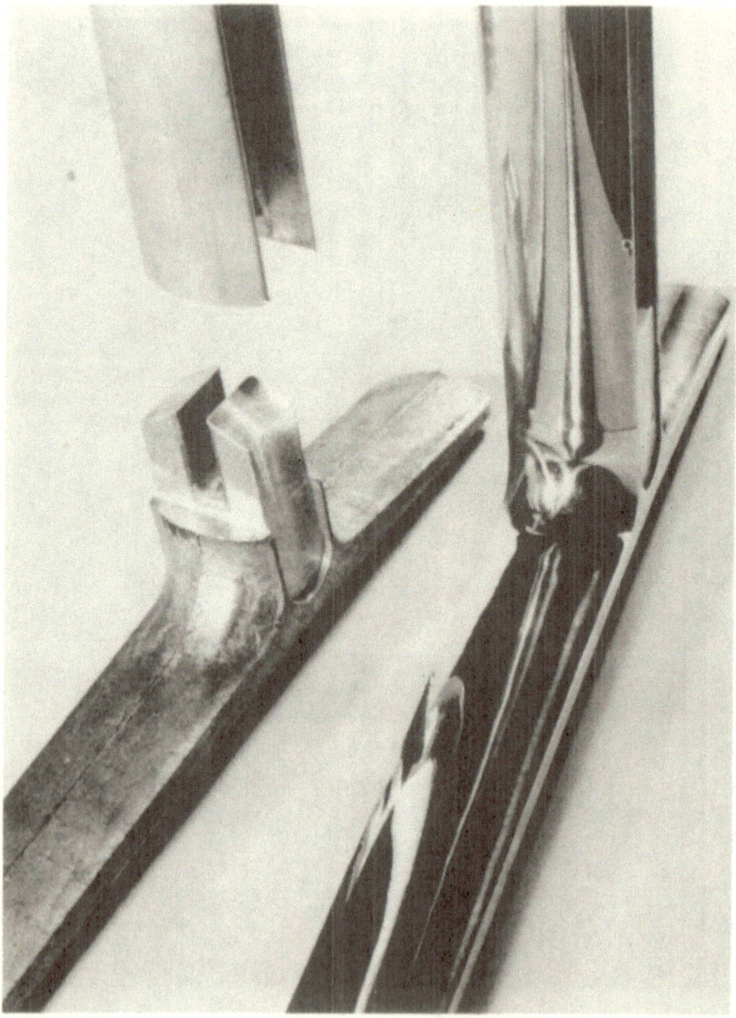

PLATE 11
Epoxy adhesive bonding the structure of chairs made by Race Furniture Limited, Sheerness, for the restaurants of the *Queen Elizabeth II* (Courtesy of CIBA-Geigy)

16

Adhesives

WHY USE ADHESIVE BONDING?

In the last few decades, there has been a very rapid development in the production of adhesives based on synthetic polymers, and these new materials have displaced many of the older adhesives derived from animal, vegetable or mineral sources. The advent of synthetic adhesives has also led to the wider use of adhesives, as the newer materials, when properly used, can produce bonds of much greater strength than has been possible in the past. In fact, it is now often advantageous to use a bonded structure rather than to use the traditional methods of joining materials such as welding, riveting and other mechanical means of fixing. This is not to say that adhesive bonding will eventually replace all of the more traditional methods, but rather to foresee a situation, probably already reached in some branches of industry, where adhesive bonding is accepted and considered on equal terms with the older methods. Clearly, the designer will, in the end, select the joining method that is most suited to the specific job under consideration. To help him in reaching that decision is one of the aims of this chapter.

What, then, are the advantages and disadvantages of bonding materials with adhesives compared with the use of other joining methods? Some principal advantages[1-3] are:

(1) Adhesives allow stresses to be distributed over wider areas, making possible lighter assemblies than could be achieved with mechanical fastenings, for example honeycomb-core sandwich panels. Examples of weight reductions that can be achieved have been given[3] as 25 per cent in compression-loaded structures and 10–15 per cent in shear-loaded structures. Local stress concentrations present in spot-welded, riveted or bolted joints are avoided in bonded materials and fatigue cracks are propagated more slowly.

(2) Joints can be designed to take advantage of the stiffening effect of bonding. Rivets and spot welds hold surfaces together only at localised points, whereas adhesives form a continuous bond along the surface. Bonded structures are therefore much stiffer and loading may be increased by possibly even 30–100 per cent before buckling occurs.
(3) The need for fixing holes, such as in rivetting, does not arise when adhesives are used. These holes can lead to weakening of the effective material cross-section.
(4) Bonding is often faster and cheaper than welding, soldering or bolting, and can be carried out at room or relatively low temperatures, thus avoiding the effect of heat on the structure being assembled.
(5) Simplified design and construction techniques can often be adopted with accompanying cost savings, and dissimilar materials that are difficult to join together by other techniques, such as concrete, stone and metal or wood, rubber and plastics, can readily be bonded.
(6) Adhesives provide joints that are airtight, impermeable to moisture, highly resistant to corrosion and chemical attack and of good electrical resistance. They can also be made electrically conducting.
(7) Fragile materials, thin films and small particles, which would be impossible to join by other methods, can be bonded with adhesives.

There are, of course, limitations to the use of adhesive bonding, chiefly:

(1) The surfaces to be bonded need to be carefully prepared.
(2) The adhesive should be handled carefully, mixed accurately and normally applied before the compounded material has gelled.
(3) Cure of the adhesive may take a long time to occur at the maximum permissible temperature, and the resulting joint may be sensitive to service at elevated temperatures. At present, adhesives cannot be used where an extended life is required at temperatures above about 300 °C.

No attempt is made here to assess the advantages and shortcomings of the many different mechanical methods of fixing, which have been summarised by many authors, such as Smerdjiev[1] and Bair[4], but it is hoped that the above observations have shown that adhesive bonding of surfaces is a valuable technique which can lead to substantial savings if correctly used, and can assist designers, architects and engineers to achieve solutions to problems that have hitherto been considered to be insoluble. The adoption of adhesive bonding complements, rather than renders obsolete, the use of traditional methods. In particular, epoxy adhesives allow strength to weight and stiffness to weight ratios to be increased markedly, making possible the design of high-strength structures that combine long service life with the ability to be lifted or accelerated with the exertion of the minimum amount of power.

CLASSIFICATION OF ADHESIVES

There are many ways of classifying adhesives, none of which is completely satisfactory. They can be grouped according to their physical type, their mode of application, their bonding requirements, the materials to be bonded for which they are preferred, and their chemical composition.

Adhesives are available as liquids, pastes, putties and solids. As solids they are usually in the form of a powder, a rod or a supported film, while the liquid form can be either a solution of the adhesive substance in a solvent or a two-phase fluid composition such as an organosol, or be 100 per cent reactive materials. Depending on their physical form, adhesives can be applied by spraying, brushing, knifing, roller and many other specialised techniques. Most, if not all, adhesives have one or more bonding requirements involving heat, pressure and time. Thus a system can cure at room temperature or require heating before the adhesive will set to an acceptable level, the major factor being the extent to which the resin is polymerised, and in certain systems an adequate degree of polymerisation can be attained only by the application of heat. Certain adhesives also require the bonding surfaces to be placed under pressure, whereas others need only contact pressure.

The most common method of classification is by chemical type, and two major groups of adhesives can be distinguished:

(1) *Inorganic adhesives.* These are based on substances such as sodium silicate and magnesium oxychloride, and are not suitable for structural bonding.

(2) *Organic adhesives.* These can be subdivided into:

(a) *Natural product adhesives.* These are based on animal or vegetable sources, for example, albumin, proteins, starches, cellulose derivatives, shellac, rosin and gums. They are used chiefly with paper, cardboard and wood, and will not be discussed here.

(b) *Synthetic adhesives.* These are based on synthetic polymers and can be grouped into three further categories:

Thermoplastic.
Elastomeric.
Thermosetting.

The names of these sub-groups indicate the physical nature of the adhesives and the type of polymer on which they are based. Thus, the thermoplastic adhesives may utilise polymers such as polyamides, polyethylenes or polyvinyl derivatives. Elastomeric adhesives can be based on rubbers such as SBR, ABS, polysulphides and silicones. Epoxy resin adhesives fall into the thermosetting group together with many others, such as those based on phenolic, urea, melamine, polyester and polyurethane resins. All of these resins need to undergo cure (cross-linking) in order to form the infusible insoluble product, which is the final form of the adhesive in the bond.

In general, thermosetting resin adhesives are more expensive than the other types, but their bond strengths are correspondingly higher and their creep strengths better. They are used most widely as structural adhesives in the assembly of load-bearing components, often in conjunction with other resins, such as an elastomer or a thermoplastic, which may be mixed with the thermoset in order to improve toughness, peel strength and impact resistance. *Table 16.1,* after Semerdjiev[1], gives a comparison of the properties of some of the common structural adhesives, the most widely used ones being the modified epoxy and phenolic systems.

PROPERTIES OF EPOXY RESIN ADHESIVES

Adhesives based on epoxy resins are widely used for bonding metals, concrete and a whole range of non-porous materials. They were first introduced in the late 1940s and established their reputation as adhesives for bonding metals in the 1950s by virtue of their ease of application and the very high strengths of the bond that they form.

Table 16.1

EXAMPLES OF COMMON STRUCTURAL ADHESIVE TYPES AND PROPERTIES (AFTER SEMERDJIEV[1])

Type of adhesive	*Forms*	*Economics*	*Working life at room temperature (ready to use)*	*Curing conditions (RT = room temperature)*	*Tensile shear strength at RT, × 10⁷ N/m²*	*Peel strength*	*Impact resistance*	*Creep resistance*	*Heat resistance, °C*	*Solvent resistance*
Phenolic–vinyl	Liq. (1 part) Liq. + powder Film	Medium	½ to 1 yr	150 °C, 20 min Pressure	1.7 – 3.5	Good	Good	Fair	100–130	Fair
Phenolic–rubber	Liq. (1 part) Film	Medium	> 1 yr	175 °C, 30 min Pressure	0.3 – 3.2	Good	Good	Good	175–230	Good
Epoxy	All forms	High	15 min to > 1 yr	RT, 24 h, 300 °C, 3 min Contact	1.5 – 4.5	Poor	Poor	Good	80–175	Good
Epoxy–polysulphide	Liq. (2 part)	High	15 min to 2 h	RT, 24 h, 65 °C, 20 min Contact	1.3 – 2.5	Medium	Medium	Medium	80	Good
Epoxy–polyamide	Liq. (2 Part)	High	30 min to 3 h	RT, 1 to 5 days, 200 °C, 3 to 5 min Contact	1.4 – 2.5	Medium	Medium	Medium	80	Medium
Epoxy–nylon	Film	High	6 h to 1 yr	175 °C, 1 h Pressure	2.5 – 4.2	Very good	Good	–	90–150	Good
Epoxy–phenolic	Liq. (1 part) Film	High	1 to 15 days	165 °C, 1 h Pressure	2	Medium	Poor	Good	260	Good
Unsaturated polyester	Liq. (3 part)	Medium	½ to 2 h	RT, 6 to 24 h Contact	1 – 3	Good	Good	Medium	80–100	Good
Polyurethane	Liq. (2 part)	High	2 to 8 h	RT, 24 h, 175 °C, 1 h Pressure	3.5	Good	Good	Good	90	Good
Cyanoacrylate	Liq. (1 part)	Very high	½ to 1 yr	RT, < 10 min Contact	1 – 2	Poor	Poor	–	80	Poor
Polyacrylate	Liq. (1 part)	Very high	1 yr	RT, 1 to 24 h, 120 °C, 1 min Contact	0.5 – 4	Good	Good	–	90–200	Good

Their use has now spread into a large number of different fields, and evidence of this widespread use is given in the later section devoted to applications (p. 240).

In addition to enjoying the general advantages of adhesive bonding over mechanical fastening, epoxy resin adhesives offer clear advantages over the other synthetic bonding materials:

(1) Epoxy adhesives do not normally contain solvents and are 100 per cent reactive. They cure without the evolution of water or other by-products, which makes them eminently suitable for bonding non-porous surfaces, and removes any concern over solvent removal, bubbling or porosity of the bond.
(2) Only contact pressure between the parts is required during cure, and cure can take place at either ambient or elevated temperatures, depending on the formulation of the adhesive.
(3) Epoxy resins cure with only negligible shrinkage, compared with polyesters and acrylics, which minimises the strain built into the glue line.
(4) The strength properties of the epoxy resins are outstanding. They offer good adhesion to a wide variety of substrates, and when properly cured have such a high cohesive strength that, when taken with their adhesive strength, failure in the adherents usually occurs before failure in the epoxy resin or at the interface. In addition, they have low creep properties and are resistant to moisture and chemical attack.
(5) The properties of epoxy resin adhesives can be varied over a very wide range by (a) using different resin–curing agent combinations, (b) the addition of fillers, plasticisers, flexibilisers and diluents to the formulation, and (c) the incorporation of a second resin, such as a phenolic or nylon, into the adhesive composition. By these techniques, epoxy resin adhesives can also be produced in many different physical forms, such as liquids, pastes, putties, powders, rods and supported films.

SELECTION OF THE ADHESIVE: PRACTICAL HINTS ON USING EPOXY ADHESIVES

There are very many commercially available epoxy adhesive products and it is often found that several different epoxy compositions can be used to bond the surfaces under consideration. Selection of the most suitable adhesive then becomes dependent on price and the handling and final properties of the adhesive. Factors such as the viscosity of the mixture of resin and hardener, the cure schedule required, the range of operating temperatures of the cured adhesive and the average strength of the joint at these operating temperatures will all be important factors in the ultimate choice.

The strength of the adhesive bond is dependent on two factors, the adhesion of the adhesive to the materials being bonded and the cohesion of the adhesive itself. Bond strengths are little influenced by glue line thickness (at least over the range 0.005–0.04 in), but are dependent upon:

(1) The materials being bonded.
(2) Surface preparation.
(3) Formulation of the adhesive.
(4) Cure schedule.
(5) Temperature of the test.

Some important points to remember when selecting and using an epoxy adhesive have been given by CIBA-Geigy [5]:

(1) It is essential to inspect the materials to be joined and make sure they are clean and free from grease. Abrading or etching of the surface of the joint is recommended.
(2) Choose an adhesive with a cure time and temperature that are practicable. For example, a hot-setting adhesive that requires a cure temperature of 140 °C and above could not be used to bond wood, because wood carbonises at this temperature, and a cold- or warm-setting adhesive should therefore be used.
(3) Cold-setting adhesives give excellent bond strengths when cured at room temperature. A higher cure temperature, even a few degrees above room temperature, will, however, increase their bond strength.
(4) If the bonded assembly is designed to operate at temperatures above 100 °C, then choose a hot-setting adhesive. The strength of a joint, once bonded, becomes less as its temperature is increased. Joints bonded with hot-setting adhesives have the greatest resistance to heat.
(5) Select an adhesive with a suitable viscosity. Adhesives of low viscosity are the simplest to apply and are recommended for large, closely fitting joint surfaces. Adhesives of high viscosity

are more suitable when application is to vertical or overhead surfaces, or when the gap along the joint is of irregular width.

(6) The thickness of the adhesive layer in the joint should be between 0.001 and 0.008 in, but it may be greater if a paste adhesive is used.

(7) When solvent-containing adhesives are used, the solvent must be allowed to evaporate before the joint is assembled. This does not apply so strictly when one or both of the surfaces is porous, as the solvent is then absorbed.

(8) Design the joint to take a shear or tensile stress, not a peel stress.

(9) The relative coefficients of expansion of the materials and the adhesive may be important if the adhesive is hot setting or the bonded assembly is designed to operate over a range of temperatures. Stresses in joints can be minimised by matching coefficients of expansion as far as possible. The use of glass fabric carriers minimises the stresses set up in the joint during cure. Typical carriers, such as a single layer of open scrim or plain-weave glass-cloth, improve peel strength, give better strength retention above the HDT of the adhesive and tend to reduce the effect of stresses that result from the bonding of materials with dissimilar coefficients of expansion.

(10) It is preferable to use an adhesive with a low modulus of elasticity if the bonded assembly is designed to withstand vibration or heavy impacts.

(11) The strength of a composite joint will be that of its weakest part. For example, aluminium alloy bonded to steel will fail at 2000–3000 p.s.i. (14–21 MN/mm^2), whereas aluminium alloy bonded to polythene will fail at 1000 p.s.i. (7 MN/mm^2) because the degree of adhesion even to treated polythene is less.

APPLICATIONS OF EPOXY ADHESIVES

In every sphere of industrial activity, and especially in aircraft, automobile, marine, electrical, electronic and civil engineering, epoxy adhesives are commonly used. Designers and architects have become fully aware of the outstanding properties of these materials and specify them regularly. They are also used increasingly in the home, the workshop and in museums and artists studios. A number of these uses are considered elsewhere. Thus the numerous applications for epoxy adhesives in building and civil engineering are discussed in Chapter 13, in the home, workshop and studio in Chapter 17, and in the electrical and electronics industries in Chapters 8 and 10. This section deals with the remaining fields of use for these remarkably versatile adhesives and the examples considered have been chosen chiefly to illustrate this versatility and the many different applications that are possible.

Metal bonding

Probably the best account of the way in which the epoxy adhesive bonding of metals can lead to significant economies over conventional joining methods, has been given by Semerdjiev[1], and much of the following discussion draws heavily from this source.

There are various mechanical means for attaching a load-carrying stud to a surface: using a nut, screwing on, rivetting or using a pin. The simplest method is bonding, and by using either a room-temperature or heat-curing epoxy adhesive, failure stresses of 14 kN/mm^2 or 25–28 kN/mm^2, respectively, can be obtained with a cylindrical stud 12 mm in diameter bonded 25 mm into the base plate. The use of an epoxy adhesive to bond a bearing bush on to a lever rather than the bush being pressed and secured with a screw has brought about economies in time and cost. The need for close-tolerance machining is eliminated, resulting in a reduction of production costs, an increase in productivity and a reduction in reject rates.

A good example of the way in which bonding can lead to savings in raw materials and machining is the general case of an axle with a journal. If the axle is made from solid material (example I in *Figure*

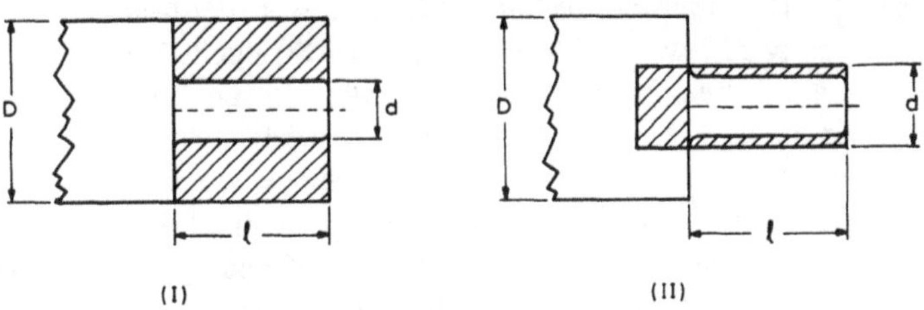

Figure 16.1. Examples of bonding of a journal to an axle

16.1), the following utilisation of material and machining factors apply for different sizes of the axle:

Diameter of axle, D, mm	*Utilisation of material,* per cent	*Machining factor,* per cent
40	39	61
50	25	75
60	17.5	82.5

where the material utilisation factor is the ratio between the weight of the finished and that of the semi-finished part, in this case $100\,(d/D)^2$ per cent.

If the journal is made separately and bonded to the axle (example 11 in *Figure 16.1*), a 28-mm diameter steel bar can be used. In this case, the ratios given above become:

Diameters, mm		*Utilisation of material,* per cent	*Machining factor,* per cent
(a) For the axle:			
$D = 40$	$d_1 = 28$	51	49
$D = 50$	$d_1 = 28$	68	32
$D = 60$	$d_1 = 28$	78	22
(b) For the journal:			
$d = 25,\ d_1 = 28$		82	18

Other relevant dimensions in the bonded example were bore depth $l = 40$ mm and bond area $A = 3500\ \text{mm}^2$. The data show that significant improvements in material use and reductions in the amount of machining required are effected when an epoxy adhesive is used rather than making the journal from the solid bar by machining.

Other examples in the mechanical engineering field abound. Ceramic and metal cutting tools can be bonded to holders, cutter heads, etc., using an epoxy adhesive which cures between 100 and 200 °C[6]. Adequate bond strengths are attained that will withstand the cutting forces, and the heat resistance of the adhesive needs to be high enough for the bond to be unaffected by the heat developed during machining.

Complicated assemblies can be made by joining simple die castings together by adhesive bonding[7]. The reject rate of a pump manufacturer was reduced from 25 per cent to nearly zero by using this method[8]. Originally, the pump part was cast as a single piece, but owing to the complexity of its interior shape many blowholes often occurred, which led to the high reject rates. Casting simple shapes and joining them together with an epoxy adhesive eliminated the need for lugs, bolt holes and flanges required for mechanical fastening and drastically reduced the reject rate.

Epoxy adhesives have also been used successfully to bond and seal header tanks to tubes in equipment such as tubular heat exchangers and oil separators. The traditional method of securing the tubes to the headers is by arc welding or soldering, but bonding has been found to be a more reliable method[7,9,10].

An outstanding example, taken from the light engineering industry, of the way in which epoxy adhesives allow not only a cheaper article to be produced, but also a stronger, better designed and more saleable item, is that of the Valor domestic paraffin heater[11]. Formerly, the paraffin reservoir was made from two basic parts, comprising a corrosion-resistant brass base to which was soldered a tinplate upper half. The assembly was later painted and the soldered joint sealed on the inside so as to prevent leaks. In recent years, Valor have found it increasingly difficult to obtain skilled labour for the soldering operation and it was undesirable to work with fluxes that had to be washed off or neutralised. As the proportion of female labour increased, Valor realised that the manufacturing process had to be changed in order to eliminate the soldering operation. The corrosion problem was overcome by the use of pre-painted steel, which also had other advantages in that it was cheaper and a thicker gauge could be used, thus making the tank stronger. As the tank was pre-painted, no further painting was required, and by means of a mechanical and adhesive joint, the tank was joined and the joint sealed in one operation. This was achieved by applying the adhesive to the rims of both halves, then assembling and clinching the seam. The tank could then be passed through a tunnel oven, hardening the adhesive without any complicated method of locating and holding the two halves together. The adhesive also provided a guaranteed seal throughout the joint. At the same time, it was possible to attach other auxiliary items to the tank by means of the adhesive. For example, a guide for the spindle used in raising and lowering the wick is assembled on the tank with adhesive before it passes through the tunnel oven. This and all other parts which are added to the tank are thus bonded in one operation, so that after the article has passed through the tunnel oven virtually all assembly operations have been completed.

As a result of redesigning the manufacturing process, Valor have eliminated soldering and painting by the use of adhesives and pre-painted steel. The assembly operation is simplified because it is carried out entirely within one area of the factory and all the adhesive joints are cured during one pass through the tunnel oven. Unskilled labour can be used because the adhesive is automatically dispensed and the application and assembly of parts which have already been prepared and laid out, are simple processes. Valor estimate that although theoretically the process is costing less because soldering and painting have been eliminated and unskilled labour is now employed, the main advantage is that they are now producing a stronger, more reliable and hence more competitive product at no greater cost to themselves. Thus by cooperation with British Steel Corporation and CIBA-Geigy Ltd. (the raw material and adhesive suppliers, respectively), Valor have designed a process suited to their requirements.

Another example in the light engineering field is the model locomotive chassis[12]. This was zinc die-cast and had two rectangular Alnico magnets as inserts, and joining these two materials by ordinary welding or brazing is out of the question. The conventional approach would be silver soldering at a very high cost and would require the services of skilled operatives at high rates of pay. There would also be the risk of burning through the zinc die-cast and/or causing disorientation and embrittlement of the magnets, and reject rates could be extremely high. In this situation, an epoxy adhesive can give a much better result at a lower cost. The basic requirement was for a high shear strength but with no peeling stresses involved. Resistance to lubricating oil, vibration and abrasion was also required. The epoxy adhesive met all of these requirements, allowing the chassis to be assembled at a lower cost with fewer rejects.

A problem in automobile engineering is the fluttering and rattling of engine bonnets and boot lids at high speeds. In order to overcome this problem, some manufacturers bond reinforcing members to the undersides of these relatively large double-curvature panels[13-15] using a modified epoxy resin adhesive. This increased stiffening not only eliminates flutter, but also allows the weight of the assembly to be decreased. In other fields of transportation, epoxy adhesives have been used more widely than in automobiles. Examples are the securing of stiffeners to decks and skins of aluminium boats and, also in the marine world, Professor Picard's bathyscaphe was bonded together with a heat-cured epoxy compound[1]. Underground railway tracks have been secured directly to the smooth floor of the tunnel by epoxy adhesive bonding, so that the tracks did not require ties or setting in a gravel bed.

Moreover, considerable cost savings were possible because the height of the tunnel could be made 30–35 cm lower than usual[16].

Semerdjiev[1] reported the development of a 'wheel-axle' set of the future by the Deutsche Bundesbahn. The aim was to reduce weight and high internal stresses and to eliminate ringing noises, disadvantages which the conventional steel set possesses. In the design prototype a hollow axle fabricated from a number of thin-walled tubes placed and bonded inside one another is used. The forged steel journals are also bonded into the axle shaft. The wheels consist of laminated epoxy–glass-fibre shell on to the rim of which is bonded a steel rim as the bearing surface. The wheels are bonded to the axle shaft. The whole set weighs 500 kg and has a load-carrying capacity of 20 000 kg compared with a conventional set of the same load-carrying capacity which weighs 1230 kg. There would therefore be a 60 per cent saving in weight, if the new wheel set were to be introduced into service.

Also in the transportation field, the most prolific user of epoxy adhesives has been the aerospace industry. Very many large modern aircraft use epoxy adhesives for joining metal parts instead of, or in addition to, riveting and brazing. Metallic honeycomb sandwich structures are also widely used in aircraft, and epoxy adhesives are employed in the production of the honeycomb panels as well as in the joining of the panels in the aircraft itself. The U.S. Air Force B-58 Hustler is reported[17] to have 85 per cent of its load-carrying exterior metal cladding adhesively bonded, much of it with an epoxy adhesive. The aircraft as a whole contains over 400 kg of adhesive, replacing 500 000 rivets[1]. The Fokker F27 Friendship and F28 Fellowship airliners also employ large amounts of adhesives in primary load-bearing and secondary structures. Epoxy-bonded honeycomb panels, possessing high strength to weight ratios together with good thermal and vibrational resistance, are widely used in aircraft for fuselage doors, side panels, floors, elevators, rudders and trim tabs. Helicopters also use honeycomb components for pod assemblies, floor frames, loading ramps and the rotor blades themselves. Substitution of a bonded blade in place of a rivetted blade increased the service life from 90 to 1200 h[1].

Epoxy adhesives and bonded honeycombs are frequently used in space vehicles and satellites[18]. An extensive account of one such use has been given[19] for the U.K. satellite Ariel-3. Launched in May, 1967, and designed to circle the earth once every 95.5 min at a distance of 494 km at the perigee and 600 km at the apogee, the satellite weighed 86 kg.

Many items of equipment in the satellite employed epoxy resins, for example the frames and structures of instruments for the various experimental packages.

The use of epoxy adhesives in the electronics industry has already been considered in Chapter 10. One 'engineering type' use of adhesives by the industry consists of the manufacture of distortion-free radio cabinets[20]. Originally, the cabinets required 150 spot welds, and consistently failed in pressure cycle tests which simulated the thermal expansion undergone by the cabinet. No failures were experienced when the weldings were replaced with a continuous adhesive-bonded joint. The same organisation used an epoxy adhesive to seal an aluminium power supply box to withstand an air-tightness test of 25 p.s.i. (175 kN/m^2). The use of the adhesive was less expensive and more reliable than welding and eliminated the need for separate seals and gaskets. Overall production costs were cut by 50 per cent, fabrication to closer tolerances was possible and the product appearance was improved, compared with a welded construction.

A different application, but one also related in an indirect sense to the radio industry, is the use of epoxy adhesives in the fabrication of the giant radio-telescope at Westerbork in The Netherlands[21]. The telescope, which became operational in June, 1970, has many essential elements bonded with an epoxy product. The equipment consists of twelve parabolic dishes, each of 25 m diameter; ten dishes, spaced 144 m from each other, are stationary and two are mobile, being mounted on a 330 m long track connected to the stationary antennae. The adhesives were used during construction and assembly to make the reflectors, to connect the reflectors to their supporting structures and to connect gear rings to their sub-structures.

The parabolic reflectors consist of about 100 segments, each having a galvanised steel supporting framework for the reflecting stainless-steel gauze. If the gauze was secured to the frame by clamps and screws, the uniformity of the reflectors would be affected, the point by point fixing causing uneven tensions. Adhesive bonding overcame the difficulties and also provided a degree of self-tensioning for the gauze as it shrank during cooling after the hot cure of the epoxy adhesive.

The twenty large gear rings, each with a diameter of 9 m, were fabricated by a novel method. Gear sections were made in the normal way by machining, and then bent and attached to a welded sub-structure by bonding with epoxy adhesive. The ends of each segment were also welded together for additional security.

Bonding dissimilar materials

The selection of examples to illustrate the versatility of epoxy adhesives for bonding dissimilar materials is difficult, in view of the very many possibilities from which one can draw. Examples of bonding concrete to metal, bauxite to asphalt, and concrete, glass, ceramics or wood to concrete, have all been quoted in previous chapters. To this list can be added other examples such as the bonding of paint brush bristles to their handles, abrasive grit to the metal disc of grinding wheels[22,23], polyester laminates to themselves and polypropylene handles to the wooden shafts of spades[24], and a PTFE sleeve to a bronze plug valve body[25]. Phenolic centrifugal brake linings have been bonded on to the steel brake shoes with greater ease than rivetting and, in sports equipment, the sycamore centres of discuses have been bonded to the metal rims[26]. Other examples of epoxy bonding in the manufacture of sports equipment are the joining of the aluminium alloy shafts of javelins to the steel points, the fixing of ferrules and other parts to glass-fibre–polyester laminated fishing rods, glueing heads to the metal shafts of golf clubs, bonding different granites together in modern curling stones, and fabricating modern skis[26,27].

Other miscellaneous uses may be mentioned, for example the manufacture of trombones and other wind instruments, of modern breathing apparatus[28] and repairing a broken stalactite.

REFERENCES

1. SEMERDJIEV, S., *Metal to Metal Adhesive Bonding,* Business Books, London (1970)
2. SKEIST, I., *Handbook of Adhesives,* Reinhold, New York, Chapter 1 (1962)
3. *How to Design for Araldite Bonding,* CIBA-Geigy Ltd
4. BAIR, F. H., in PATRICK, R. L. (Editor), *Treatise on Adhesion and Adhesives,* Part 2, Marcel Dekker, New York, Chapter 7 (1969)
5. *How to Use Araldite Adhesives,* CIBA-Geigy Ltd.
6. EHRLICH, W., *Feltig. Tech. Betr.,* **10,** No. 8, 459 (1960)
7. WICK, C. H., *Machinery, N.Y.,* 74, No. 2, 87 (1967); **74,** No. 3, 73 (1967); 74, No. 4, 76 (1967); **74,** No. 5, 64 (1968)
8. HESS, E. F., *Precis. Metal Mold.,* **21,** No. 6, 44 (1963)
9. *CIBA-Geigy Technical Notes,* No. 255
10. *CIBA Aspekte,* (July 1965)
11. GOLDING, J. H., *Assembly Fastner Engng,* **9,** No. 2, 25 (1971)
12. BAKER, A., *Industrial Fastening and Assembly Conference,* London (1967), Paper 12

13. *Araldite Epoxy Resin Adhesives,* CIBA Publication No. A41a
14. *Metal Wkg. Prod.,* **106,** No. 41, 72 (1957)
15, VAN SCIVER, H. D., *U.S. Pat.* 3, 385, 744.
16. GEYS, K., *Bauengineur,* **39,** No. 3, 94 (1964)
17. BONDARUK, W., *Aviat. Age,* **28,** No. 3, 72 (1957)
18. HILLESLAND, H. L., *Sampe J.,* 63 (April-May, 1968)
19. *CIBA Products in the Ariel-3 Satellite, CIBA Technical Notes,* (Nov. 1967)
20. *Metal Prog.* **93,** No. 5, 108 (1968)
21. *Araldite in a Giant Radio-Telescope, CIBA-Geigy Technical Notes,* No. 10, (1971)
22. *Araldite in Grinding Wheel Technology, CIBA-Geigy Technical Notes,* (March, 1970)
23. *Metal Prog.,* **93,** No. 5, 111 (1968)
24. *CIBA Technical Notes,* (Jan. 1968)
25. SCHMIDT, J. E., MATTHEWS, A. L. and HORNBLOWER, W. J., *Prod. Engng.* **34,** No. 11, 43 (1963)
26. *CIBA Resins in Sports Equipment,* Part 1
27. *CIBA Resins in Sports Equipment,* Part 2
28. *Araldite in Breathing Apparatus, CIBA Technical Notes,* (Sept. 1966)

17
Arts and Crafts; Antiquities and the Home

In this chapter are considered some of the non-industrial uses for epoxy resins in which large-scale consumption of the resins does not occur, although turnovers and profit percentages can be high to the seller of the epoxy-formulated product. Rather, these uses are of interest in showing how epoxy resins can assist in a very valuable social and cultural field, namely the creation and particularly the preservation of works of art and craft and antiquities.

The latter part of the chapter also considers how epoxy-based products are being used in many homes by the handyman and do-it-yourself enthusiast, and underlines the versatility of these materials. All of these uses depend on the outstanding adhesive properties of epoxy resin systems, together with the good mechanical properties that can be obtained. Naturally, the innumerable ways of modifying a basic epoxy formulation and the various application methods that are possible, e.g. impregnation, laminating, casting, spraying, spreading and brushing, all contribute to the extreme versatility of the systems and their consequent wide use.

PRESERVATION AND RESTORATION OF WORKS OF ART AND CRAFT, AND ANTIQUITIES

The process of restoration carried out in museums consists in assembling broken fragments, rebuilding missing parts, reinforcing fragile parts, assembling the whole and removing dirt and foreign matter. In this work, synthetic resins are playing an increasing part owing to their superior properties compared with those of the natural products used hitherto, such as beeswax, animal glue and plaster. Within the group of synthetic resins, epoxy resins have a special place, by virtue of:

(1) The wide range of physical and mechanical properties that can be obtained by varying formulations without the need to incorporate solvents.
(2) Their ability to cure at room temperature with very little shrinkage and no pressure being required to form a bond, which is virtually free from internal stresses.
(3) Their outstanding adhesion to most substrates, including metals, stone, porcelain, glass, wood and paper, all of which are common materials met regularly by the restorer.

Epoxy resin formulations are used by restorers and museum laboratories chiefly as adhesives and filling compositions, replacing the solvent-containing adhesives for repair work where other means of jointing, such as screwing or brazing of metals, is not possible because of disfigurement or the effect of the local heating occasioned by welding and brazing.

Epoxy adhesives have apparently proved particularly successful for bonding smooth, non-porous surfaces of small area in relation to the size and weight of the object under repair[1]. These adhesives provide much stronger joints than the other adhesives used in this work, and can be permanently coloured with colouring pastes so that a glue line or section being filled or rebuilt can be made to blend exactly with the original object. The wide use and versatility of resin systems are best illustrated by considering some examples of work in which they have been used. Much of this account of the world-wide use of epoxy resins in restoration work is drawn from the excellent *Technical Notes* published by CIBA-Geigy Ltd., to whom the author is grateful for permission to quote extracts.

Bonding, filling and rebuilding work

The joining together of glass fragments is an excellent example of the use of epoxy adhesives. Ancient glass objets d'art, such as a Roman crystal goblet, and a 'Calandarium Perpetuum' bonded on glass, have been restored[1,2]. The goblet had the missing part of its base replaced with a suitably coloured epoxy system, and the calandarium was re-assembled from the 100 or more pieces into which it had been shattered, using an epoxy adhesive to join the fragments of glass together.

Epoxy resin adhesives have also been widely used to repair

porcelain articles[3]. Today, antique porcelain is becoming rare, especially sound pieces. Moreover, porcelain is particularly susceptible to damage and fragments can be irretrievably lost. Restoration work aimed at bringing the object back to its original appearance has, in the past, used mechanical retention such as banding, binding, fastening with nuts and bolts and rivetting. In comparison with these methods, an epoxy-bonded joint is much less visible to the eye and also ensures that stresses are distributed throughout the joint rather than being concentrated in the neighbourhood of a mechanical fixing. Repairs with epoxy adhesive are therefore extremely strong and, if the article is broken again, failure often occurs in other parts of the porcelain and not in the joint.

Metal objects can be repaired particularly well by using a metal-filled epoxy formulation. This is especially so for iron objects unearthed during excavations, as the colour of a metal-filled epoxy formulation corresponds very closely with the colour of a cleaned iron object such as a spear head or iron helmet. After curing, the epoxy adhesive can be given a metallic lustre by polishing it with steel-wool. The low mechanical strength of badly corroded metal objects often presents serious difficulties to the restorer. On the one hand, the products of corrosion which so mar the shape and appearance of the specimen must be removed but, on the other hand, they may be harder than or at least as strong as the specimen itself. The use of epoxy resin to reinforce the reverse side of the object ensures that it can be safely handled during restoration work. This technique was employed in the preservation of the Dead Sea scrolls[4,5].

The repair and reinforcement of stone objects and detached wall paintings can also be readily effected with an epoxy formulation, which is coloured to match the surrounding stone and contains powdered stone as a filler. The strength and adhesion of the resins are sufficient to support heavy fragments and, if the cohesive strength of the stone is insufficient, this too can be remedied by suitable treatment.

An example of this type of work is provided by a section of an interior wattle-and-daub wall left from the ancient 'Haus zum Königstahl' in Zurich, which was demolished in 1938. The section, weighing several hundred kilograms, was decorated with 'secco-a-colla' (distemper) paintings on both sides. Through the numerous faults and cracks, the inner structure of vertical spruce laths with interwoven hazel could be seen. This framework carried the mortar rendering which was the substrate for the paintings. At the outset, the whole structure was reinforced and the hollow interior filled with an epoxy composition containing appropriate fillers (limestone, white clay, sand and zinc oxide). The filled structure was then fixed to the studs again with an epoxy adhesive, and the whole was re-erected.

An example of the restoration of a stone object is the repair of terracotta bas reliefs by Jan van Logteren, dated 1734. This Dutch artist, the son and pupil of Ignatius van Logteren (1685–1732), has other works in Haarlem Cathedral in The Netherlands and at Woburn Abbey in the U.K. The reliefs were broken into several pieces, and these have been joined and made good with epoxy resins.

Epoxy resins are also being used in a new technique, developed by the National Gallery in London, for the transfer of frescoes to a new mounting. In the past, when frescoes had to be removed from their original position because of damage to the supporting wall, it was customary to use a multi-stage process by which the paintings were transferred to a new plaster or cement mounting. In order for the new support to have the same stiffness as the original wall, its weight had to be enormous. In the new process, an epoxy resin composition is used in conjunction with a paper honeycomb core material, and has been used successfully in the case of the Domenchino frescoes[6]. The process has the advantage that no special walls or hanging arrangements need to be provided for the display of the paintings, and the frescoes are better able to be transported. In a similar way, this technique can be used for mounting mosaic pavements, rather than using plaster, cement or concrete backings[7].

Epoxy resin products of very low viscosity can be used to preserve timber that has become infested with wood worm. Waxes are often used for this application as well as solvent-containing paints, solutions of glues and other materials. However, epoxy resin systems have proved markedly superior because of their low viscosity, the absence of solvents and their good wetting properties. Nevertheless, additional holes need to be bored in large wooden objects in order to obtain sufficiently deep penetration. A wooden statue of Bhudda from Japan has been restored by this technique at the National Workshop in Kyoto. Gilded wooden carved panels in the Abbey of Muri, Aargau, Switzerland, had been damaged by woodworm and some parts had been completely destroyed and were held together by the chalk base coat. However, a repair was effected by the following method. Silicone rubber was placed over the gilding and the fronts of the panels were embedded in polyurethane foam. The damaged wood and loose wood flour was then removed from the chalk backing, which was then given a thin coating of an epoxy resin, the 'mould' thus formed being filled with 'plastic wood'. Another example of the repair of wooden objects is the restoration by the Department of Conservation of the Victoria and Albert Museum, London, of a French marquetry cabinet. Some of the legs of the cabinet had been attacked by woodworm and the affected parts were cleaned away and made good with an epoxy material.

Copies and casts

It is often necessary to prepare copies of museum specimens and, in the past, plaster of Paris has been the material most commonly used for this purpose. Nowadays, epoxy resins are being used on an increasing scale for copies, which are usually taken from moulds made in polysulphide or silicone rubber. The epoxy copies are dimensionally stable, unbreakable and weatherproof, which certainly cannot be claimed for plaster copies. The mould contours are reproduced with complete accuracy and the epoxy copies can be metallised or electrolytically coated. In fact, the restorer can reproduce faithfully almost any material, owing to the wide range of epoxy resin systems, fillers and colourants available. An example is the copy made of a Roman sandstone relief by the Museum at Augst near Basle, Switzerland. Copies have also been made of Late Roman silver from Kaiseraugst, Aargau, Switzerland. In particular, an Achilles plate with a diameter of 53 cm and a statuette of Venus are both excellent examples of the technique of coating the epoxy copy with a layer of metal by electroforming while still in the mould. The most intricate details of the originals have been particularly well produced in the finished copies.

THE PRESERVATION OF MONUMENTS

Probably the most impressive use of epoxy resins to preserve monuments was in the saving of the Abu Simbel temples in Egypt[8].

These two rock temples, built in about 1250 B.C., constitute a work of art of major importance, and were liable to be submerged by the flood water which would be created by the new Aswan dam.

The rescue operation mounted by UNESCO in 1960 following an international appeal consisted in cutting both temples bodily out of the sandstone rock, dividing them into sections each of not more than 30 tons in weight and moving them to a site 200 ft higher and above the future water level. The work began in 1964 and the temples were cut into 3000 numbered blocks. Epoxy resin mortars were then used to anchor the 7-ft steel bolts in holes bored in the sandstone rocks, by which the sections were lifted. The resins were also injected into cracks to reinforce and strengthen the porous rock, and painted on the decorated surfaces of the sections so that no material would break away. The work is now completed and there are times, when the waters of the lake are at their highest, when the aspect of the monuments differs little from that before the salvage operations began. Furthermore, twice a year, in mid-February and mid-October, the sun penetrates to the sanctuary of the great temple and shines on the figures of the god Amon and of Rameses as it did by the design of the ancient Egyptians over 3200 years ago.

In Japan, epoxy resins have been used for some years to preserve ancient monuments such as the bronze statue of Bhudda from a temple at Nara known as the 'Gakko Bosatsu' and weighing about 2 tons. Epoxy resin adhesives were used to secure the head of the statue and to fill numerous cracks that had been caused by earthquake damage and fire since the statue was made in 688. In the ancient city of Madurai, in the South of India, a mediaeval bronze statue was also repaired with an epoxy adhesive. The feet of the statue had completely broken off and a repair was made using an epoxy system to replace the sand core which supported the bronze outer shell of the statue.

NEW ART AND CRAFT FORMS

The fine arts, painting and especially sculpture, possess many examples of the use of epoxy resins in the creation of new works. In painting epoxy resins have been used as the binder for paintings where thick impastos have held objects on to the surface of the

PLATE 12
Epoxy bonded window panels by Miss Eugenia Politis at the Royal Trafalgar Hotel, London (Courtesy of CIBA-Geigy)

PLATE 13
The new Roman Catholic Cathedral at Liverpool, showing the stained glass bonded with an epoxy adhesive (Courtesy of Shell)

canvas or other ground. In a similar way, collage and low relief work can utilise epoxy compounds as a means of holding together the objects that form the work.

However, it is in sculpture and modelling that epoxy resins are more widely used, and many examples can be cited[9,10]. Epoxy adhesives can be used to join parts of a sculpture together and a good example of this use stands in Norrkoeping in Sweden[11]. This town, about 100 miles south-west of Stockholm, possesses a most unusual piece of monumental sculpture, entitled 'Prisma', which was created by Vicke Lindstrand. This obelisk stands 37 ft high, weighs 33 tons and consists of 3500 pieces of glass held together with an epoxy adhesive. The resulting structure is as strong as reinforced concrete and rests on a base weighing 55 tons. A particularly attractive aspect of the sculpture is the optical effect obtained. The glass is sea-green and the refraction from the very thin glue lines cannot be distinguished from the glass itself. The sculpture changes colour constantly according to the lighting of its surroundings.

When filled usually with a metal powder, an epoxy system can be applied directly on to an armature as a knifing putty. When the epoxy system has hardened, the forms can be refined by using a knife and file and a final effect obtained with sanding cloth or steel-wool. Burnishing with a clean piece of metal before the resin has completely hardened is also an effective way of finishing the sculpture.

A further technique in which epoxy resins can be used is known to sculptors as the 'cold cast metal technique'. This process, in which polyester or epoxy resins can be used, is a means of producing a metal-surfaced casting without the need to use foundry-cast bronze. The method has been well described, in conjunction with polyesters[12], and consists in first taking a plaster mould of the clay master. The plaster negative is then sealed and coated with a release agent. Using this prepared plaster mould, an epoxy cast can be made by two slightly different methods. The first, and most popular method, is to cast each half of the mould separately and then to join the halves together. An epoxy gel çoat filled with metal powder is brushed carefully on to the two mould surfaces. When the gel coat has cured, a second laminated resin layer is layed up on the gel coat as a reinforcement. When both halves have cured, they are joined together using more filled epoxy gel coat, with glass-cloth being used for reinforcement across the joint. The second method is to place the gel coat, and then the laminated backing, directly on to the inside of the complete mould. In both methods, the plaster is removed by soaking in water and chipping away the softened plaster.

An artist who uses this technique is the well known Brazilian sculptor Ilie Gilbert. Previously working mainly in bronze, Gilbert now uses filled epoxy compositions for the creation of his portrait heads[13]. The American artist Frank Gallo has also used epoxy resins in this way for many of his life-sized sculptured torsos[10].

For centuries, coloured glasses have been shaped and made into mosaics to form highly decorative windows, especially for churches. The method used to make these stained glass windows has remained essentially unchanged for many hundreds of years, until recently, when epoxy-based compositions have been used to obtain novel and beautiful effects. Traditionally, the pieces of glass are joined together by means of H-shaped strips of lead (known as cames), soldered at the joints, and the panel made waterproof by rubbing putty under the lead strips. The window then requires strengthening when in place in the building, which is achieved by cementing iron bars as unobtrusively as possible to the mullions of the windows. The lead cames can often be restrictive to the artist, and the use of epoxy adhesives has provided for the first time a means of removing the restriction of the cames. Epoxy adhesives can make a perfect bond between glass and glass, as they do not require exposure to air in order to harden, nor do they give off volatile compounds during cure. Moreover, the thin bond line required is completely colourless and transparent. To make a window without cames, the separate pieces of stained glass are glued on to a sheet of plate glass by means of a clear epoxy adhesive. The stained glass used is relatively thin and is the modern equivalent of antique glass, and is cut accurately to shape so that no white light can shine between them. Numerous examples of windows made by this technique (which is known as glass appliqué) are quoted in a CIBA-Geigy publication[14].

Probably the most impressive examples are the installation of stained glass windows into the octagonal lantern of Blackburn Cathedral[15], where a total area of glass of nearly 1000 ft^2 was used, weighing nearly 2½ tons, and the windows for the 75 ft lantern of Liverpool Roman Catholic Cathedral[16], which required about 1 ton of epoxy adhesive (plate 13).

A further method of making stained glass windows involves the use of cast slab glass about 1 in thick. This thick glass is a modern introduction and has opened up new possibilities for the artist. Thus, the glass slabs may be chipped and flaked at the edges, creating a brilliance comparable with that of fine jewellery. In making the window, the pieces of glass are first cut and chipped to shape for the desired pattern, and are laid out within a frame the size of the window. Concrete may then be poured around the pieces, but most artists now prefer to use instead an epoxy resin formulation. The epoxy formulation has the advantage of being much lighter in weight, and therefore easier to handle, and provides a much more powerful bond to glass than does concrete and the window panel is therefore stronger. With concrete cames, it is necessary to cast the cames to the full thickness of the glass, perhaps 1 in thick. When using epoxy resins cames, thicknesses of only ½–¾ in are needed, and thus the edges of the glass can be left fully exposed. This allows sparkling and iridescent effects of great beauty to be obtained, especially when the edges of the glass have been chipped.

EPOXY RESINS IN THE HOME

There are two main types of epoxy product used by the home handyman and hobby enthusiast: epoxy adhesives and epoxy stopping compounds. Epoxy laminating with glass-fibre is also carried out, but for the most part this field is dominated by polyesters.

Epoxy adhesives are available in many countries, usually as a twin-tube pack. One tube usually contains the resin, lightly filled, and the other the curing agent, which almost always is a polyamide. The mixing ratio is not critical and there is little or no skin sensitisation effect from the polyamide. Cure takes place at room temperature although heat assists the rate of hardening. A recent development has been the introduction of very rapidly curing epoxy adhesives in which a polymercaptan is used as the curing agent. These versatile adhesives can be used to bond all manner of items in and around the house – broken china, toys of metal or wood, knife blades to handles, concrete ornaments and other items, garden tools, hooks to garden walls, etc. These adhesives also find innumerable uses in the maintenance and repair of cars, caravans and boats.

Epoxy stopping compounds or putties are also widely available as a metal, silica or micro-balloon filled composition. The metal-filled putty is used to fill holes in car bodies, exhaust pipes, petrol tanks, cylinder blocks, to repair metal drain pipes, water or oil tanks and in general to fill holes or dents in metal articles. The silica-filled material can be used as a sealant for the surrounds of baths, sinks, basins and also externally around windows and to fill joints between wood and steel or concrete. The micro-balloon type is known as plastic wood and behaves in an identical fashion to wood when shaped or sanded with woodworking tools.

REFERENCES

1. *Araldite in the Restoration of Antiques,* CIBA Technical Literature
2. *CIBA Aspekte,* Sondernummer (1966)
3. *Araldite in the Repair of Porcelain and Glass,* CIBA-Geigy (1971)
4. *The Dead Sea Scrolls, CIBA Blätter,* No. 149, 20 (1957)
5. BAKER, H. W., *Engineering, Lond.,* **181,** No. 4701, 194 (1956)
6. *The Domenchino Frescoes, CIBA Technical Notes,* No. 245 (1963)
7. *Araldite in the Restoration of Roman Mosaics,* CIBA (1965)
8. *Araldite in the Saving of Abu Simbel,* CIBA (1966)
9. NEWMAN, T. R., *Plastics as an Art Form,* Pitman, London (1965)
10. ROUKES, N., *Sculpture in Plastics,* Watson-Guptill, New York (1968)
11. *CIBA Technical Notes,* (April, 1968)
12. PERCY, H. M., *New Materials in Sculpture,* Tiranti, London (1970)
13. *CIBA Technical Notes,* (Nov., 1970)
14. *Araldite in Stained Glass Windows,* CIBA (1966)
15. HAYWARD, J., *Stained Glass Windows,* CIBA (1968)
16. GIBBERD, F. *et. al., Building,* **112,** No. 6474, 107 (1967)

18

Production Engineering

One of the earliest uses for epoxy resins was as a tooling material in the aircraft and motor-car industries, and this still remains an important, well established application for the resins. Not only is the epoxy resin used as a tooling material itself, that is as a working tool, but also as an auxiliary or secondary tool. Nowadays, epoxy tools of various types are used in many different industries such as the foundry, ship-building, boat-building, pottery and domestic equipment industries and light engineering in general in many countries of the world.

It is now clearly established that the use of epoxy tooling resins in place of the more traditional tooling materials results in faster, more efficient and lower cost tooling-up. The use of the resins has not introduced any new approach to engineering practice, but rather has brought about large savings in time and money and provided a tool which often has a performance superior to that of one made in conventional materials.

Epoxy tools can be broadly divided into two types:

(1) Working tools where the tool actually shapes or forms metal or other material.
(2) Auxiliary or secondary tools used as aids to the primary engineering processes.

The former type include moulds for forming plastics and for the production of glass-reinforced laminates, and drop hammer and press tool faces. These tools are usually made by casting the resin composition into a solid shape or by using the resin to form a facing surface around a heavily filled core. The latter type of tool consists of drilling and welding jigs and checking fixtures and prototype and master models for product design, models for die sinking and Keller machines. These are usually laminated structures built from epoxy resins and glass-fibre, often with epoxy laminated tubing as stiffening.

The manufacture of tools of both types uses fairly simple casting and hand lay-up laminating techniques, the liquid epoxy system being laid against an existing shape and cured in position to form a hard infusible tool without machining. This demonstrates one of the primary advantages of epoxy tooling, namely that the epoxy tool costs less because of the ease with which it is formed.

THE PROS AND CONS OF EPOXY TOOLING

Whether a tool should be made from sheet, bar or rod metal, or from an epoxy composition, depends upon a number of factors, the chief ones being:

(1) *Raw material and manufacturing costs.* Epoxy material costs are high compared with those for metals. In contrast, the manufacturing costs of epoxy tools are considerably lower than those for metal tools. The epoxy systems are liquids or at least compositions that flow and can easily conform to both simple and highly complicated shapes. Therefore, the more complex the working surface of the tool, the greater is the advantage of using an epoxy formulation. Metals, on the other hand, are normally used for tools with flat or regular surfaces, as they require expensive machining and hand finishing if they are to have complex contours. Thus the higher raw material cost of the epoxy formulation is greatly outweighed by savings in construction time and labour costs. Complex-contour epoxy tools are less than half the price of similar metal tools and delivery times are cut to a quarter[1].

(2) *Manufacturing method.* The liquid epoxy formulation is cast or laminated against a model or mould and often cures at room temperature. No subsequent machining to shape is required. Metal tools for complex contoured shapes require expensive machining operations, fettling and finishing after forming, bending, welding, etc. This machining requires the skilled operation of expensive machine tools. Epoxy tools require no expensive equipment for their manufacture and take fewer man hours to carry out the simple operations involved.

(3) *Reproducibility.* Epoxy tool surfaces have the exact reverse image of the master model or mould profile, whatever its complexity. Any number of identical accurate epoxy tools can be reproduced from one accurate model. With metal tools, the reproducibility of surface contours is dependent on the

Table 18.1

STRENGTH COMPARISONS OF TOOLING MATERIALS (COURTESY CIBA-GEIGY LTD[1])

Property	*Epoxy–glass-fibre laminate*		*High-tensile steel*	*Mild steel*	*Cast iron*	*Aluminium alloy*	*Titanium alloy*
	*A**	*B**					
Density, lb/ft³	100	120	485	485	435	175	280
g/cm³	1.6	1.9	7.8	7.8	7.0	2.8	4.5
Ultimate tensile strength, lbf/in²	17 000	57 000	125 000	66 000	26 000	60 000	125 000
MN/m²	119	399	875	462	182	42	875
Young's modulus, 10^6 × lbf/in²	1.1	3.3	29.0	30.0	14.0	10.0	16.0
GN/m²	7.7	23	203	210	98	70	112
Specific Properties (relative to cast iron):							
Strength to weight ratio†	3.5	5.4	1.9	1.4	1	3.8	3.4
Stiffness to weight ratio†	1.9	2.3	1.2	1.2	1	2.2	1.6

* A = hand lay-up material, 50 per cent resin by weight; B = pressure-moulded woven rovings, 30 per cent resin by weight.

† For beams in simple bending and of equal width and weight per unit length.

$$\text{Strength} \propto \frac{\sqrt{\text{ultimate tensile strength}}}{\text{density}}; \qquad \text{Stiffness} \propto \frac{\sqrt[3]{\text{Young's Modulus}}}{\text{density}}$$

accuracy of machining during the manufacture of each individual tool.

(4) *Specific strength and stiffness. Table 18.1* shows that on a weight for weight comparison, an epoxy laminate is superior to high tensile and mild steel, cast iron and titanium alloy in both specific strength and specific stiffness. Therefore, although an epoxy laminate with a greater thickness must be used in order to obtain the same strength and rigidity as steel, even at this greater thickness the laminate is lighter in weight. However, tool steels have the higher tensile strength, hardness and rigidity essential for press tools to be used for long production runs. Epoxy press tools are often made with metal inserts to carry out shearing.

(5) *Weight.* The specific gravities and volume per unit weight of a number of metals and epoxy systems are given in *Table 18.2*. The low weight of epoxy tools has important advantages on the shop floor. Lifting tackle is not required for jigs, fixtures and models and, in fact, the light weight of the epoxy tool encourages its frequent use. Saving in storage space is also possible as more epoxy units than metal units can be stored in a given area before floor loading limits are exceeded. For drop hammers, on the contrary, light weight is a disadvantage and the weights of these epoxy tools are increased by casting the working surface in epoxy resin around an undersized metal core.

Table 18.2

SPECIFIC GRAVITIES AND VOLUMES PER UNIT WEIGHT OF EPOXY SYSTEMS AND METALS

Property	*Epoxy laminate*	*Epoxy casting, 200 per cent filler*	*Steel*	*Zinc alloy*	*Aluminium alloy*	*Lead*
Specific gravity	1.6–1.9	1.8	7.8	6.8	2.8	11.3
Cubic inches per lb	14½–17	15½	3½	4	10	2½

(6) *Thermal properties.* Epoxy resins cannot be used for the direct forging and casting of metals at temperatures above the thermal decomposition point of the resins. The high thermal conduct-

ivity of metals is necessary in tooling for plastics where the tools are subjected to rapid heat cycling over a large temperature range. The low thermal conductivity of epoxy resins is useful in conserving heat in plastic sheet while it is being thermoformed.

(7) *Storage stability.* Epoxy tools are unaffected by changes in atmospheric conditions which would alter the size and accuracy of wooden tools and patterns. Epoxy tools may even be stored in the open. In contrast, metal tools must normally be stored under controlled conditions as a precaution against corrosion. Cleaning corroded tools can be very expensive, especially in labour costs.

(8) *Chemical resistance.* Metal tools are attacked by many chemicals that have no effect on epoxy resins, and they are also susceptible to atmospheric corrosion. Epoxy tools are unaffected by moisture and are resistant to attack by chemicals and solvents that are likely to be encountered during tooling processes.

(9) *Modification and repair.* Rapid re-profiling of epoxy tools is possible because of the good adhesive properties of the liquid epoxy system and of the easy machineability of the cured resin. Re-machining of metal tools is much more time consuming and the addition of metal for re-profiling is normally impossible.

(10) *Mould materials.* Metal tools are cast in moulds, usually sand or metal, that can withstand the melting temperatures of the alloys being cast. Epoxy tools can be cast or fabricated in moulds made from a wide range of materials, because high temperatures are not involved.

TYPES OF EPOXY TOOLS

Epoxy tools are used at all stages of manufacture, from the initial design of the product, through prototype production, to full-scale production and final assembly. The epoxy tools used at these various stages are considered in the following sections.

Master models for product design

At the product design stage, the design drawings are followed by the building of a three-dimensional model in plaster, clay, plastic or metal. This master is then used to make several highly accurate duplicate models in epoxy resin. These duplicate models can be produced rapidly and cheaply, and can be sent to a number of departments in the firm such as design, tooling and planning, as well as to sub-contractors, to supplement the detailed drawings. Using epoxy plastic wood models, the time required for car-body development programmes has been reduced to a third, with great savings in tooling man-hours and a speeding up of pre-production schedules.

The U.S. aircraft industry is now making master models for its new planes by using extrudable epoxy syntactic foams which, when cured, can be automatically machined and milled[2]. It is claimed that on average this technique has cut the time required to make master models by 50 per cent and permitted cost savings of up to 55 per cent compared with the previous template and master models. Instead of laying on plaster by hand, the resin is extruded on to a frame, allowed to harden and then machined to the correct shape by a numerically controlled machine tool. McDonnell-Douglas, a major user of the new technique for the DC-10 aircraft, has used 1 in thick aluminium honeycomb panels faced on both sides with epoxy–glass laminates as the supporting base and also as the shaping support structure for the foam. The panels are built into an egg-crate type of structure having a cell size of perhaps 20–30 in. These cells are then filled with polyurethane foam, and when this has set the epoxy foam is extruded on top. Boeing have built models as large as 22 x 30 ft using this method. The advantages of the technique are: faster production of models; longer life from the tool – hopefully 20 years; and fewer damaged tools through bad handling and attack by moisture. Several profiles can be obtained from the same sub structure. If the shape changes during the development of an aircraft the master model can be redesigned by either adding or machining off material. The cost of this operation with a plaster model would be prohibitive, and it would be preferable to prepare another model.

Master models for metal tooling

Epoxy compounds have been used with great success to make Keller and other die models. These models can be cast rapidly and cheaply from the original master with no machining needed. The epoxy model thus made can withstand the wear of the Keller duplicator while maintaining exact dimensions. The epoxy model is also much lighter than a wooden or metal model, has superior dimensional stability and is unaffected by ageing and moisture. A further advantage of the epoxy model is its versatility, as it can be used not only as a Keller model but also as a spotting aid for the final finishing of tools and as a checking fixture for the pressings made from the tools. This results in large savings in labour costs for tool manufacture and a marked reduction in the delay between design and production.

Metal shaping tools

Although inferior in strength and toughness to die steel, epoxy tooling resins are available that possess strength and toughness sufficient to shape and form not only the softer metals such as aluminium, but also some steel sections. Although the life of epoxy metal-forming tools is short compared with that of die steel tools, in many instances it is preferable and more economical to use an epoxy tool for a short-run job rather than invest in a high-quality steel tool that has a longer life but which will be used only for a small production run. Epoxy press tools and drop hammers are therefore particularly favoured for prototype and short- and minimum-run production, especially when the dies have complex contours. To machine such a die from tool steel is costly; to cast the same die from epoxy resin will result in a considerable reduction in labour costs and construction time.

Although the aircraft industry has been the chief user of epoxy press tools, drop hammer dies, stretch dies and draw dies because of the short-run nature of its production, the motor-car industry is also an important user of epoxy tools of all types. Production runs in the car industry are usually much longer than the life of an epoxy die and epoxy dies are used for prototype and certain short-run work. However, the term short-run is relative and resin dies have been used successfully for runs of up to 100 000 items. Naturally, the life of the tool depends upon the type of material being formed and the severity of the forming operation.

In some interesting research, British Leyland[3] has examined the suitability of alternative methods of producing body panels for the four 'problem areas' of car-body panel production:

(1) Prototypes, where numbers of up to 50 are produced by skilled panel beaters.

(2) Pre-production panels in numbers up to 1000 and where production tools cannot be made available.

(3) Low-volume production panels for sports and prestige cars, usually in numbers from 5000 to 10 000.
(4) Variant forms of high-volume production vehicles such as estate cars and vans, usually in numbers from 10 000 to 50 000.

The costs of production per panel for these four situations were investigated, using different production methods, and the results are represented in *Figures 18.1* to *18.4.* Actual costs are not given in the

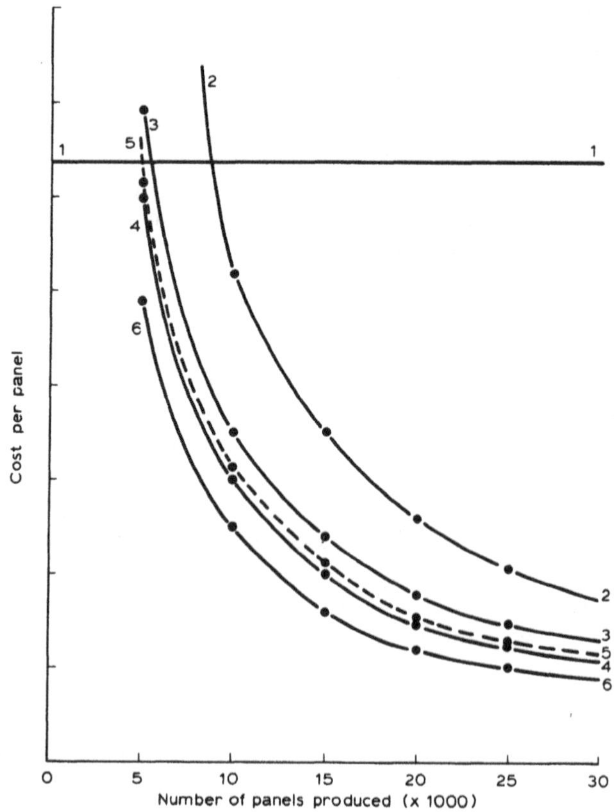

Figure 18.1. Cost per panel for producing prototype panels: Line 1 represents the cost of the conventional method of production using skilled panel beaters, and the cost of producing panels by other means are shown as a percentage of conventional production costs. Line 2 represents cost per panel when using Jabroc tooling, line 3 when using Kirksite tooling, line 4 when using epoxy resin tooling, line 5 when using a casting technique with Prestaform low melting point alloy and one casting frame, and line 6 when using Prestaform with two casting frames. (From *Mach. Prod. Engng.*, **121**, No. 3134, 834 (1972))

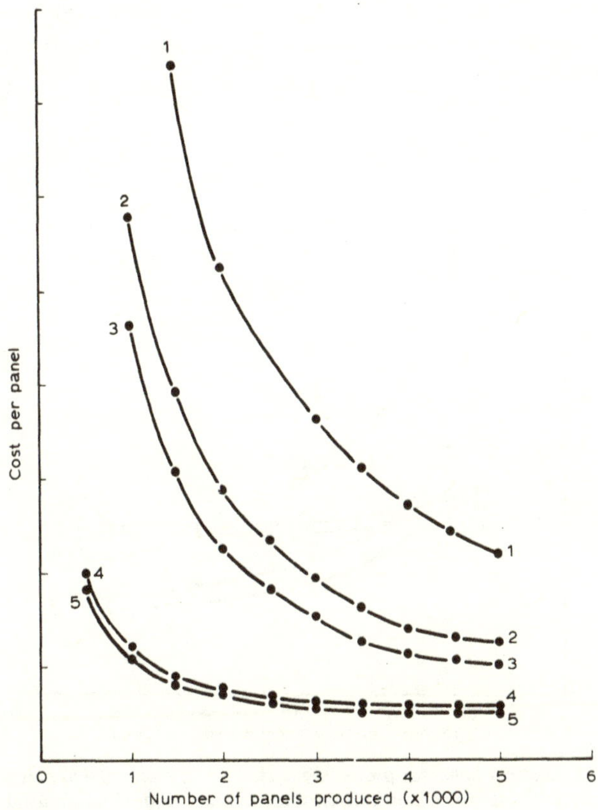

Figure 18.2. Cost per panel for producing between 500 and 5000 panels: Line 1 represents the conventional method of production using cast-iron dies. Line 2 shows the cost per panel when using Kirksite, line 3 when using epoxy resin tooling, line 4 when using a combination of a Kirksite die and a rubber punch, and line 5 gives the cost per panel when using an epoxy resin die and a rubber punch. (From *Mach. Prod. Engng.*, **121**, No. 3134, 834 (1972))

graphs, but the relative savings achieved by the various methods are immediately apparent. The costs were based on average outer-body panels and were limited to the costs incurred in the drawing operations. The results of this costing exercise show that epoxy resin press tools could be beneficial in each of the four problem areas considered.

Epoxy dies can also be made in a relatively short time and can therefore be used in an emergency to keep a production line running while a steel die is being completed or repaired. Resin dies can also shorten the time taken to switch production from one model of car to another and therefore greatly increase the competitive position of the manufacturer.

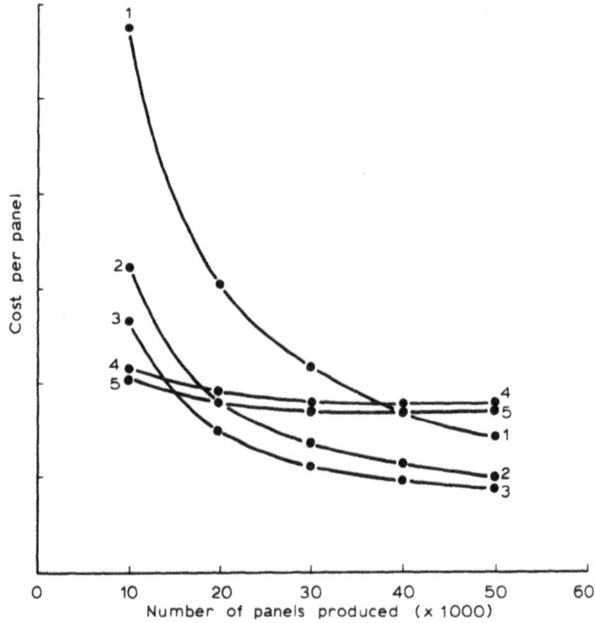

Figure 18.3. Cost per panel for producing between 5000 and 50 000 panels: Line 1 represents the cost of the conventional method of production using cast-iron tooling, whereas lines 2 and 3 represent the costs for producing panels when employing Kirksite tooling and epoxy resin tooling, respectively. Line 4 indicates the cost per panel when using the rubber pressing technique in conjunction with a Kirksite die, and line 5 shows the cost when employing a rubber punch and an epoxy resin die. (From *Mach. Prod. Engng.*, **121**, No. 3134, 834 (1972))

Moulds for thermoforming plastics

The vacuum forming of plastics and rubbers requires the pre-heated sheet of these materials to be drawn by vacuum over the tool so that it takes up the exact profile of the tool. Epoxy vacuum tools operate at temperatures up to 155–160 °C and produce mouldings of quality comparable with those from an aluminium alloy tool. A comparison of epoxy resin with aluminium alloy vacuum forming tools has been given by CIBA-Geigy[1] (*Table 18.3*).

Table 18.3

COMPARISON OF EPOXY RESIN AND ALUMINIUM ALLOY VACUUM FORMING TOOLS (COURTESY CIBA-GEIGY LTD.[1])

Epoxy Resin Tools	*Aluminium Alloy Tools*
Cast to true size, shrinkage during and after curing is negligible	Cast oversize to allow for shrinkage.
Machining and hand finishing is nil or minimal	Expensive machining is normally essential
Lower thermal conductivity of epoxy reduces heat transfer from sheet to tool. Resultant heat conservation assists forming, reduces pre-heating temperatures	Higher conductivity assists rapid heat cycling in tool, but the cycle time is not appreciably reduced as pre-heat temperatures are normally greater to offset rapid heat transfer on sheet to tool contact
Higher thermal capacity ensures more uniform heat distribution during moulding	Cold spots liable where sheet is in final contact with tool
Easy positioning of cooling tubes directly behind working surface	Tools normally drilled to accommodate tubes. Not practical directly behind tool surface
Modification of tool simple to carry out owing to ease of machining epoxy systems and of building up contours with additional resin	Modification usually possible only where machining is involved. Building up contours by addition of metal is normally impossible

Injection moulds for thermoplastics can also be fabricated from epoxy compositions and offer two advantages over steel moulds:

(1) The ability to produce cheaply prototype components for evaluation from resin moulds before embarking on the manufacture of expensive production tools in metal.
(2) Economic short-run production of components can be carried out with resin moulds where metal moulds would be prohibitively expensive.

Contact and matched moulds

Most epoxy and polyester composites are made by the contact moulding technique. This requires a master mould to be produced, which is then used to make an epoxy production mould for the composite. Thus, for a small capital outlay on tooling-up with epoxy moulds, large complex mouldings such as boat hulls and vehicle bodies can be produced with ease and at low cost. Compared with moulds constructed in polyester resin, epoxy moulds have better dimensional accuracy and stability, greater surface hardness, longer life and improved dimensional stability under working conditions and in storage. They are also cheaper than metal moulds and less time consuming to construct especially when the mould surface is complex. Matched mould tooling in epoxy resins is also used for low-pressure moulding of epoxy and polyester resins and in particular for dough moulding.

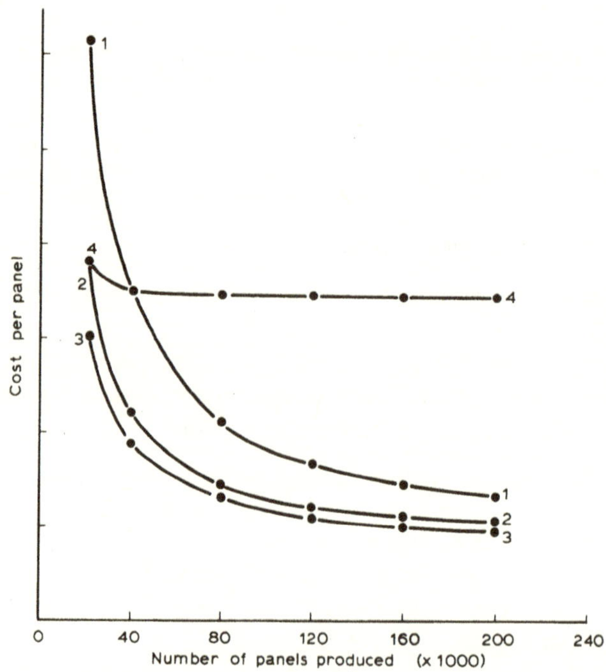

Figure 18.4. Relative costs when producing panels in numbers between 20 000 and 200 000: The conventional method of producing such numbers is by means of cast-iron tooling, the cost per panel for which is indicated by line 1. Line 2 shows the cost per panel when employing Kirksite tooling, line 3 when using epoxy resin tooling, and line 4 when using a rubber punch and Kirksite die. (From *Mach. Prod. Engng.*, **121,** No. 3134, 834 (1972))

Tooling aids

Production engineering requires various jigs, fixtures and checking aids to ensure accurate positioning of workpieces and to check the dimensions, shape and position of the finished article. These jigs and fixtures have, in the past, been made from metal, wood and plaster, but these materials have now mostly been superseded by epoxy resins. Epoxy resins are regarded as the supreme material for constructing these aids, because of the strength, lightness, ease of fabrication, dimensional stability and resistance to rough handling, chemicals, greases, etc., of the epoxy tools. This latter point can result in significant cost savings. Skeist[4] reported that the Douglas Aircraft Corporation spent $50 000 per year at a single plant to refurbish metal tools. Moreover, bulky epoxy tools can be stored outdoors, which is not normally the case with metal tools.

The use of epoxy tools has also made some tools portable for the first time. Previously, lifting tackle may have been required in order to move them around, but when fabricated in epoxy resin they may be 70–80 per cent lighter. An example of this great difference in weight is the jig used by MAC Engineering Ltd.[1] in the production of tubular steel field gates. The weight of the epoxy welding jig is 60 lb whereas the weight of a similar jig in steel is 600 lb, and handling the epoxy jig is therefore very much easier than manoeuvering the steel jig. The epoxy jig was also constructed in 90 per cent of the time it took to make the metal jig and a further unforeseen advantage was that the weld spatter which adhered to the metal jigs fell off the epoxy laminated tubing of the plastic jig.

Foundry patterns and core boxes

The foundry industry, like the toolmaking part of the engineering industry, makes extensive use of the technique of 'reproduction moulding', in which an accurate master model is first made in wood, clay, plaster or other material and this master is then used to reproduce a number of identical duplicates. These duplicates are used for a variety of purposes, including the moulding of the final, finished article.

Traditionally, the foundry industry has used wooden and occasionally metal patterns, mounted on wooden or steel plates, in its production processes. However, wooden patterns, although acceptable as originals or masters, are not very satisfactory from the point of view of dimensional accuracy when they are in continual use. On the other hand, metal patterns are not always economically acceptable, especially for short- and medium-length production runs. Wooden patterns, while satisfactory for use in hand moulding and demoulding processes, are not adequate in modern moulding machines especially when jolt–squeeze or high-pressure methods are used. The higher the quality requirements of the finished casting, the greater are the disadvantages of wooden patterns. In particular, wood undergoes dimensional changes resulting from its hygroscopic nature and there is some loss in accuracy with each release of the pattern from the mould.

Important aims of modern foundries are:

(1) To obtain greater dimensional accuracy of castings through the establishment of closer tolerances and at the same time to guarantee consistency of dimensions for longer production runs.
(2) To achieve economic production of mould components combined with high quality.
(3) To achieve shorter production times and greater reliability in meeting delivery deadlines.

Epoxy resin patterns and fittings have helped the industry to attain these objectives because of the following main advantages they have over traditional wooden equipment[5]:

greater hardness;
excellent dimensional stability;
can be removed from the mould without vibration;
when stored for long periods and used for repeat orders, suffer no dimensional changes;
easily built up and repaired, allowing design changes to be introduced easily and quickly;
checks on wall thickness can be carried out in the pattern shop so that trial castings are unnecessary, thus saving a considerable amount of time;
the release properties of the casting can also be checked in the pattern shop.

These advantages of epoxy patterns over wooden patterns also establish them as excellent alternatives to metal patterns and equipment, which suffer from long delivery times, high tooling costs and the inability to be modified rapidly. A striking example of the use of epoxy patterns and core boxes as interim equipment was in the manufacture of the 2.7 litre V6 Essex engine by the Ford Motor Company[6]. When production of the engine was due to begin, unforeseen delays in manufacture held up delivery of the metal foundry equipment needed to make the intake manifold. A serious delay in the production of the new engine might have been caused, but sets of epoxy patterns and core boxes were rapidly made and production was started on time using the epoxy equipment. Castings were produced at the rate of 80 per hour over a period of 6 months until the metal parts could be delivered.

Table 18.4 shows the savings that a Canadian foundry achieved by using epoxy patterns and core boxes on five different jobs[7]. It is interesting to note that on four of the five tools the material costs of the epoxy tool were greater than those of the metal tool, but the labour saving in every case greatly outweighed the added material cost. In addition, the speed with which the plastic tools could be produced meant that they were ready for use much sooner than the comparable metal tools.

Table 18.4

COST COMPARISON BETWEEN METAL AND EPOXY FOUNDRY TOOLS (AFTER BURTON[7])
All costs in Canadian dollars*

Tool	*Metal*		*Epoxy*		*Saving*
	Material costs	*Direct labour,* h	*Material costs*	*Direct labour,* h	
Side frame pattern	280.00 (aluminium)	600	488.00	150	1142.00
Side frame corebox	39.00	60	70.00	11	116.00
Corebox B	17.00	31	30.00	9	53.00
Corebox C	10.00	16	14.00	6	26.00
One intricate pattern	100.80 (brass)	450	11.20	50	1289.00
				Total	2626.00

*Labour rate taken at the 1957 rate of $3.00/h. Master pattern costs not included.

Another example of the savings that can be made by using epoxy resins was that of International Harvester[8]. The time for a model change on a frame casting was cut from 3 weeks to 1 week and the total cost reduced from \$700 to \$300 by modifying the existing pattern with epoxy resin. In a separate case, the cost of a design change in a cylinder head was reduced from \$3500, the cost of a new pattern, to \$1500, which is the cost of modifying the existing pattern with epoxy resin.

REFERENCES

1. *Araldite Epoxy Resins in Tooling*, CIBA-Geigy Ltd.
2. Mod. Plast. International, 2, No. 5, 52 (1972)
3. *Mach. Prod. Engng*, **121**, No. 3134, 834 (1972)
4. SKEIST, I., *Epoxy Resins*, Reinhold, New York, 109 (1958)
5. *Araldite in Modern Foundry Practice*, CIBA-Geigy Ltd.
6. *Araldite Tooling by the Ford Motor Company*, CIBA-Geigy Ltd.
7. BURTON, H. A., *Trans. Am. Foundym Soc.*, **65**, No. 432, 36 (1957)
8. *Steel*, **140**, 108 (Feb., 1957)

Index

Adhesives, 233–248
  advantages, 233
  applications, 240
    bonding in new structures, 191–196
    bonding reinforcements, 190
    in building industry, 184–198
      remedial uses, 185–191
    in the home, 259
    metal bonding, 241
    bonding dissimilar materials, 247
  bulk mechanical properties, 196
  classification, 235
  creep tendencies, 196
  elastomeric, 236
  fire resistance, 195
  hints on use, 239
  limitations, 234
  inorganic, 235
  miscellaneous bonding applications, 197
  natural product, 236
  organic, 235
  properties, 236
  selection, 238
  synthetic, 236
  thermoplastic, 236
  thermosetting, 236
Aerospace industry,
  use of adhesives in, 245
  use of composites in, 221
Aircraft industry, use of composites in, 222
Amine-cured epoxy paints, 34, 39
Antiquities, restoration and repair, 249
Anti-tracking additive, 124
Araldite type I, 9
Armatures, epoxy powder protection, 105
Art works, preservation and restoration, 249–253
Automobile bodies, use of adhesives in construction, 244

Bonding
  in new structures, 191–196
  reinforcements, 190
Bridge decks, expansion gap nosings, 181
Bridges, epoxy bonding, 192
  surfacing, 174–182

Cable jointing compounds, 117
Can linings, 49, 50
Casts of museum specimens, 253
Ceramic tiling, epoxy grouts, 202
Ceramics, restoration and repair, 251
Circuit-breaker, 109, 110
Coal-tar paints, 30, 44
Coatings, *see* Marine coatings; Paints, anti-corrosive; Pipes, steel; Powder; Surface coatings.
Commercial background, 11–14
Composites, epoxy resin-based, 207–232
  applications, 217–229
  costs, 214
  designing with, 214–216
  fibrous reinforcements, 211–216
  future for, 228
  limiting factors, 216
  matrix resins for, 209
  pipe-lines, 219
  properties, 208, 214
Concrete,
  bonding to concrete with epoxy adhesives, 188
  crack repairs, 185
  moulds, epoxy laminates for, 202
  resin-based, 203
Containers, protective painting, 50

Corrosion,
anti-corrosive paints, 29–52
costs, 33
Cure accelerators, 7
Curing, 4, 5
agents, 9, 10
Cycloaliphatic resins, 10
non-tracking, 124

Dielectric properties of epoxy resin and porcelain, 125
Diepoxides, development, 8
Diluents, 6
Diphenylolpropane, 8
Distribution switchgear, insulators, 107–114
Drums, interior coating materials, 51

Effluent pipelines, internal coatings, 68
Electrical,
industry, use of composites in, 221
insulation, 87–134
properties of epoxy resins, 21, 91
Electronic application of epoxy resins, 135–151
encapsulating techniques, 138–145
printed circuits, 145–151
Electronics industry, use of adhesives in, 246
Electropainting, 49
Encapsulation in resin, 135
polyesters for, 136
polyurethanes for, 137
selection of polymer for components, 136
techniques,
liquid resin casting, 138
liquid transfer moulding, 144
machine dispensing, 138
powder coating, 145
pre-form moulding, 145
transfer moulding, 139
Engineering,
production tooling, 261
use of composites in, 227
Epichlorohydrin, 9
Epon, 9
Epoxy ester paints, 34, 38
Epoxy laminates, *see* Laminates
Epoxy powder coatings, 72–86
advantages, 73
costs compared with epoxy stoving enamel, 79
compared with solvent-borne paint, 81
development, 75
economic analysis, 77
future uses, 85
insulation in stators and rotors, 105
limitations, 74
uses, 83
Epoxy resins,
adhesives, 233–248; *see* Adhesives
commercial background, 11–14
construction industry uses, 153–204
costs, 13
cure accelerators, 7
dielectric properties, 125
diluents, 6
fibre-reinforced, designing with, 214–216
fillers, inert, 6
fire retardants, 7
flexibilisers, 6
flooring, 157–171; *see* Flooring
formulations, 11
history of development, 8–11
insulation in rotating machines, 95–100
non-industrial uses, 249–260
properties, chemical, 23
electrical, 21, 91
mechanical, 19–22
physical, 16–19
processing, 24
reasons for use, 15–25
reinforcements, 8
resinous modifiers,
road surfacing materials, 174–182
sales in U.K., 11
technology, 4
weather-resistant, 124
Epoxy systems, properties, 24
Epoxy–bitumen road surfacing, 178
Epoxy–coal tar,
paints, 44
road surfacing, 174
Epoxy–glass laminates,
in switchgear, 113
properties, 147

Fillers, inert, 6
Fire,
resistance, 19
of adhesives, 195
retardants, 7
Flexibilisers, 6
Flooring, 157–171
cost/performance assessment, 162
domestic, 158
future developments, 170
industrial, property ratings, 161
institutional, 159
requirement ratings, 158
self-levelling, 165
terrazzo, 169
trowelled, 167
types, 164
Foundry patterns and core boxes, 273
FR-4 grade laminate, 148
Frescoe mounting, 252

G-10 grade laminate, 148
G-11 grade laminate, 149
Gas pipelines, 65
Generators, insulation, 93–106
Glass, repair with adhesives, 250
Glass–epoxy laminates
in switchgear, 113
properties, 147
Grouting
compounds, strengths, 185
in building industry, 184–198
remedial uses, 185
Grouts for levelling, 199

Heat deflection temperature, 18
History of epoxy resin development, 8–11
Hull coatings, 58

Insulating materials,
requirements of, 94
temperature classification, 95
Insulation, electrical and electronic, 87–151
for outdoor use, 121–134
uses for, 125–132
in rotating machines, 93–106
function and properties, 93

Insulators, 110
anti-tracking additives, 124
cross-arm, for overhead line structures, 129
for outdoor use, 121–134
future of, 133
glass-fibre, 127
high-voltage line, 125
in railway overhead lines, 131
low-voltage line, 128
materials other than epoxy resins, 132
weather-resistant, 124

Kerb-stones, concrete, epoxy adhesive bonding, 197

Laminates,
copper peel strength, 151
dielectric constant, 150
dissipation factor, 150
flame retardance, 150
flexural strengths, 148
for concrete moulds, 202
FR-4 grade, 148
FR-5 grade, 149
future developments, 149
G-10 grade, 148
G-11 grade, 149
higher HDT, 149
humidity resitance, 150
punchability, 151
strength retention at elevated temperatures, 150
Liquid transfer moulding, 144
Locomotive models, use of adhesives in construction, 244

Marine coatings, 55–63
decks and superstructures, 59
hulls, 59
primers, 56
Metal bonding with epoxy adhesives, 241
Mica
coil insulation, 97
thermosetting resin binder, 98
Modelling, use of epoxy resins in, 257
Monuments, preservation, 253
Motor vehicles, protective painting, 49

Motors, electrical,
insulation, 93–106
epoxy powder, 105
windings, epoxy protection, 104, 105
Moulds,
contact and matched, 271
for thermoforming plastics, 270
Museum
laboratories, use of epoxy resins in, 250
specimens, copies and casts, 253

Novolac, 9

Oil
industry, use of composites in, 217
pipelines, 67
Overhead line structures, 129

Paintings, epoxy resin binders, 256
Paints,
air-drying epoxy esters, 34, 38
anti-corrosive, 29–52
primers, 56
uses, 46–51
chlorinated rubber, 34
electrodeposition, 49
epoxy-phenolic stoving systems, 34, 44
high solids and solventless coatings, 39
marine coatings, 55–63
spray rig for solventless systems, 43
two-pack amine-cured systems, 34, 39
vinyl, 34
zinc-rich epoxy primers, 45, 56
Patents, 8, 13
Pipes, steel,
coatings, 64–71
external, 69
internal, 65–69
Polyesters for encapsulation, 136
Porcelain, repair, 250
Powder, epoxy, *see* Epoxy powder.
Printed circuits, 145–151
properties, 146, 147
Production engineering, 261–276
Properties of epoxy resins,
chemical, 23
electrical, 21
mechanical, 19–22
physical, 16–19
processing, 24

Radio-telescope, use of adhesives in construction, 246
Railway overhead line insulators, 131
Rebuilding of objets d'art, use of epoxy adhesives, 250
Reinforcers, 8
Resin
casting, 138
concrete, 203
Road
accident rate, effect of epoxy surfacing, 179, 180
marker buttons, bonding, 197
structure, epoxy bonding, 193
surfacing materials, experience in Europe, 175–181
experience in U.S.A., 174
future developments, 182
London anti-skid programme, 177
Rotating machines, insulation, 93–106
Rubber paints, chlorinated, 34

Salt water pipelines, epoxy coatings, 67
Sand consolidation with epoxy resins, 201
Sculpture,
cold cast metal technique, 257
use of epoxy resins in, 257
Sewage pipes, internal coatings, 68
Shellgrip road surfacing, 178
Ships,
anti-corrosive paints, 55–63
decks and superstructures, 59
hulls, 59
interiors, 63
primers, 56
tank linings, 61
Shuttering for concrete mounds, 202
Skid preventing surfaces, 177
Soil consolidation with epoxy resins, 201
Sports equipment, use of composites in, 227

Stained glass, use of adhesives for, 258
Steel
    pipes, coatings, 64–71
    protection by anti-corrosive paints, 46
Stopping compounds, 260
Stoving,
    epoxy powders for, 83
    finishes, 9
Surface coatings, 27–86
Switchgear,
    epoxy-glass laminates, 113
    insulation, 107–114
        future trends, 112

Tanks, epoxy linings, 48
Temperature,
    heat deflection, 18
    range, 16
Terrazzo flooring, 169
Thermal properties, 16
Tile grouts, 202
Timber preservation, 252
Tooling
    aids, 273
    resins, 261
        pros and cons, 262
Tools, epoxy,
    types, 265–276
        contact and matched moulds, 271
        foundry patterns and core boxes, 273
        master models for metal tooling, 266
        master models for product design, 265
        metal shaping tools, 267
        moulds for thermoforming plastics, 270
        tooling aids, 273
Transfer moulding, 139
Transformers, insulators, 114
Tunnel, epoxy paint coating, 48
Turbo-generators, epoxy insulation, 96

Valor heater, use of adhesive in manufacture, 243
Vinyl
    paints, 34
    wash primers, 56, 57

Water pipelines, epoxy coating, 67
Weather-resistant epoxy resins, 124
Windows, stained glass, use of adhesives for, 258
Wood, plastic, 260

Zinc-rich epoxy priming paint, 45, 56

www.ingramcontent.com/pod-product-compliance
Lightning Source LLC
LaVergne TN
LVHW101459200926
848483LV00001B/170